ENGINEERING IN A LAND-GRANT CONTEXT

ENGINEERING IN A LAND-GRANT CONTEXT

THE PAST, PRESENT AND FUTURE OF AN IDEA

Edited by Alan I Marcus

Purdue University Press
West Lafayette, Indiana

Library of Congress Cataloging-in-Publication Data

Engineering in a land-grant context : the past, present and future of an idea / edited by Alan I Marcus.

    p. cm.

 Includes bibliographical references and index.

 ISBN 1-55753-360-1

 1. Engineering--Study and teaching (Higher)--United States. 2. Technical education--United States. 3. State universities and colleges--United States. I. Marcus, Alan I., 1949-

 T73.E473 2005

 607.1'173--dc22

                  2004024926

For the faculty and students of Iowa State University's History of Technology and Science Program (HOTS) past, present, and future

◼ Contents

Acknowledgments

I would like to thank George J. McJimsey, former chair, Department of History, Iowa State University, for his financial support this venture and the members of the History of Technology and Science Program—Amy Bix, David Wilson, James Andrews, Matthew Stanley, and Bernhard Rieger—for offering their assistance and good counsel.

■ Introduction

In 1862, the U.S. Congress passed the Morrill College Act. This act permitted each state to receive the proceeds from the sale of federal lands to create an institution of higher learning within its borders. Congress took great care to designate that these new land-grant colleges catered to the "sons and daughters of farmers and mechanics" but it did not limit training to their parents' specialties. Instead, the act promised to prepare these young women and men for a myriad of occupations, the "several pursuits and professions of life."

At the heart of the new institutions rested a simple premise: American democracy depended on an educated, satisfied and socially mobile yeomanry. To supporters of the land-grant movement, established colleges, mostly on the Atlantic seaboard, had become bastions of privilege. Wealth or birth served as necessary prerequisites for entrance and these restrictions prevented the children of the industrial classes from assuming meaningful positions in society. Long-existing schools had come to serve as de facto gatekeepers to the upper classes. They both provided the veneer and connections to achieve business success and guarded admittance to law, medical and theological schools.

But that was not all. Congress's enabling act also authorized states to use a small portion of the funds to establish "experimental farms." These farms were to stand as lighthouses for the state's farmers, employing promising but unconfirmed techniques and undertaking risky work that no individual farmer could afford. Through these farms, new practices, new crops and new methods to enhance agricultural vitality would be introduced through the American countryside and the economic basis for a rural yeomanry would be preserved.

From the start, the institutions had a difficult time balancing their tripartite mission. By training youngsters in subjects not related to farming or mechanical work, were they not teaching children to deprecate and leave the farm or shop? Whose claims took precedent: those of the students or established farmers seeking to have the college investigate new methods? How active should state legislatures be in college affairs? Who determined what it was best to teach the sons and daughters of farmers and mechanics?

At least as critical was the proper relationship at each institution between the interests of mechanics and farmers. Was one group more important than the other within the state and should that translate into different institutional priorities? How exactly does one further the interests of mechanics? The law made special accommodation for experimental farms, but it offered no such guidance for shops or other forms of investigation in the mechanic arts.

In the 1860s, 1870s and 1880s, each of these issues was greeted with a sense of urgency but none was settled with a sense of finality. Most have been revisited time and again during the last 140 years, and other themes and considerations have emerged. How have land-grant universities differed from state and other public universities? What is the appropriate relationship at these colleges between the production of knowledge and the application of knowledge? Is there a tension between the production of knowledge and its dissemination? Is all knowledge equal? What areas of knowledge demand or merit special status and concentration?

For land-grant colleges and higher education in America generally, the past two decades have been a time of heightened introspection. Popularity of Total Quality Management and other management systems have compelled land-grant colleges to justify themselves and their distinctiveness. At the same time, states have demanded "accountability" as legislators attempt to justify significant state expenditures. These schools have been called upon to identify distinctive visions for themselves and clarify their central missions. In many cases, land-grant colleges have determined that a commitment to the public weal—engagement in the affairs of the state and its citizens—marks them as distinct from other American colleges and universities. But what exactly the terms of engagement entail has been left for each college to describe and define through its actions and programs.

That wave of introspection, definition and redefinition indirectly led to this volume. Some two decades ago, the State Board of Regents of Iowa capitalized on the growing prominence of Iowa State University's Department of History and its faculty in the history of science and technology to create the History of Technology and Science Program (HOTS). The HOTS program was the first PhD program in the humanities in the 125 year history of the institution and it signaled a certain maturation of the university. The university's traditional strengths in the sciences and engineering, coupled with the library's extensive collections in those areas, became the natural focus around which to engage in doctoral study in the humanities, the sina qua non of a broad-gauged modern research multiversity.

Over the past twenty years, the HOTS program has achieved national and international recognition as our research and our students' research has been integral to the framing of contemporary understandings of the history of science and technology. The program's scholarly success has been so significant, in fact, that we thought we ought to design a distinct event to commemorate its twentieth anniversary. But rather than merely celebrate the occasion, we chose to cerebrate it; we marked the anniversary by creating and sponsoring a lecture series, an activity consonant with a flourishing intellectual enterprise. And since the HOTS program is a tribute to the land-grant ideal as expressed in the later twentieth and early twenty-first centuries, we decided to investigate the history of that ideal.

We could not do justice to the entire 140 year history of the ideal in a brief lecture series so we fixed on an area where we had considerable expertise, the formulation and application of engineering education in a land-grant context. Even then we had to be selective. We chose to outline and explore three important topics and critical themes. Each of these topics and themes is locked in time. The first, integration of engineers and engineering education within the college, was particularly important in the initial half-century of land-grant college development, while the second, the forces external to the college and the state that help direct the course of engineering education, is especially appropriate in the half-century after World War I. The third, the conscious reformulation of the land-grant ideal, stands as testimony to the introspection and assessment of the last several decades.

More specifically, the first section considers how engineering educators gained standing roughly on a par with their agricultural colleagues in the initial decades of the land-grant colleges. David Harmon argues that the erection of a national forum—rather than working through the states—proved a critical maneuver to achieving success. Alan Marcus and Erik Lokensgard's essay takes the case one step further but returns the focus to the locality. They show at one school a pattern replicated at many schools during the first half of the twentieth century: the creation of interdisciplinary research to unify and to give shape to the efforts of the increasingly complex university, including engineering education, and thus to serve a wide spectrum of state interests.

The second section's focus on forces external to land-grant college engineering divisions investigates three different phenomena. Terry Reynolds discusses the distinctly American concept of accreditation of engineering schools—Europeans have governmental ministries—and finds that the nation's youngest professional engineering organization, The American Institute of Chemical Engineers, was the driving force in the standardization of curricula and credentials. Deborah Douglas concentrates on the creation of a new engineering specialty, aeronautical engineering, and demonstrates how the efforts of private foundations and the federal government help locate it within land-grant universities. Amy Bix explores how wartime situations create emergencies, a consequence of which was the large-scale introduction of women into land-grant engineering programs during World War II.

The third division is part of the introspection of the present. Howard Segal looks to the past to assess the present. Segal views contemporary plans for the reorganization of land-grant college engineering and demonstrates how a firmer knowledge of its past would ensure that its formulators were not doomed to replicate the flaws and failings of their predecessors. In the final essay, Bruce Seely extends Segal's formulation, looking to the past to plan the future. He examines curriculum development over the past century and a half and notes paths and trends in the present and likely directions for the future.

1 Engineering and Agriculture

Collegiate Conflict

Internal Dissension at Land-Grant Colleges and the Failure to Establish Engineering Experiment Stations

David L. Harmon

With the passage of the Morrill Act of 1862, the United States established land-grant colleges to benefit primarily the sons and daughters of farmers and mechanics. Specifically, the leading object of the legislation was "to teach such branches of learning as are related to agriculture and the mechanic arts, in such a manner as the legislature of a state may prescribe, in order to promote the liberal and practical education of the industrial classes in the several pursuits and professions of life." The mechanic arts, however, generally held a secondary position in the colleges and their programs. Agricultural education remained the primary concern of these institutions and further benefits from the Hatch and Adams Acts served as solid expressions of this agricultural emphasis.[1]

Friction between agricultural supporters and mechanic arts proponents continued far beyond the initial formation of land-grant schools, and it contributed to the defeat of federal legislation aimed at creating engineering experiment stations. Ultimately, land-grant colleges would be forced to appeal to state legislatures for engineering experiment station support.

Several scholars have considered the history of America's engineering experiment stations, examining their contributions to knowledge and professionalism. This study focuses on the establishment of early engineering experiment stations and examines the obstacles that academic institutions faced in advancing the engineering sciences, practice, and profession. Some historians have argued that the inability to establish engineering experiment stations resulted from the factionalism between land-grant colleges, the state universities and the federal government. While this conclusion proves correct on the surface, it does not fully represent the circumstances surrounding the existing factionalism and the impact it had on the agricultural and mechanical arts. Indeed, much of the animosity between the state university proponents and the agricultural college supporters ceased with acceptance of the Morrill Act provisions.[2]

Although factionalism between the state universities and the land-grant colleges may ultimately have contributed to the non-passage of several engineering experiment station bills, it is also essential to consider the internal politics and tensions of each group. Inside land-grant institutions themselves, internal strife and jealousies between the agricultural faction and the mechanical arts supporters directly contributed to the defeat of engineering experiment station legislation. Such problems were exacerbated by dissension and debate within the land-grant college guiding body, the Association of American Agricultural Colleges and Experiment Stations, as well as by disagreements within the newly formed Society for the Promotion of Engineering Education. These long-standing internal disputes served as the primary reason for the failure, on six separate attempts, to establish a federal endowment for engineering experiment stations. Land-grant institutions, troubled by the absence of internal uniformity and unable to fully resolve internal strife and jealousies, found it difficult to make the case for federal aid.

The push toward the establishment of engineering experiment stations clearly reflected the growing importance of mechanical laboratories to engineering studies in late-nineteenth-century America. Once established and funded, stations helped perpetuate a shift toward a more scientific approach to engineering. This change included the appearance of professional journals, full-time research professors, and curriculum development emphasizing the role of mathematics and physics in the classroom. The establishment of engineering experiment stations furthered this more scientific approach, primarily by providing upper-level and graduate students a specific arena in which to practice their knowledge. Students often built upon the work of their professors by using the engineering laboratories to test, research and apply their findings. Robert Thurston of Cornell embodied this history, as a leading engineering professor who taught and trained his students to apply these new ideas.

Robert Thurston, as Edwin Layton noted in "Mirror Image Twins: The Communities of Science and Technology in 19th-Century America," was "perhaps the foremost champion of basic research in the engineering sciences." Continually urging that engineering labs be "established in connection with engineering schools," Robert Thurston founded "two of the earliest and best-known engineering laboratories in America, at Stevens Institute of Technology and Cornell University."[3]

Thurston influenced the American engineering profession greatly. Perhaps his most significant influence, at least from an academic perspective, came through the knowledge and training he provided his students. Many of these students moved on to head engineering departments, experiment stations, and national societies themselves, all the while expounding and expanding upon their mentors' scientific research techniques and ideas. For instance, George Bissell and Anson Marston advanced the ideas of their mentor at Iowa State College and proved the driving force behind the establishment in 1904 of the first state-sponsored station—the Iowa State College Engineering Experiment Station.[4]

Starting in 1846, when Iowa joined the Union, supporters of a state university and traditional methods of study acted to halt or reverse legislative support for recognizing agricultural education. Organized on December 28, 1853, the State Agricultural Society of Iowa served as the mouthpiece for agricultural college advocates to fight back. Supporters not only promoted the value of agricultural education, they also urged Iowa to create forms of education more "in line with the trend of a democratizing and industrializing nation." In an 1854 *Farmer and Horticulturist* article, "State University—Scientific and Agricultural School," George F. Magoun went as far as to suggest the state establish "a Polytechnic School, like that of Paris," claiming the chief remaining need for education was a technological content and method.

Magoun and his fellow supporters of an agricultural college met with sizable resistance from other Iowans. Often caught in quarrelsome debates, agricultural college supporters in the state legislature continually found themselves conceding to their opponents' demands. Even after March 1858, when the Iowa Legislature and Governor provided for the establishment of a "State Agricultural College and Farm," critics continued to agitate for repeal of the legislation.

Similar conflicts ensued between state universities and agricultural institutions around the nation. For instance, in Pennsylvania, an institution had been formally organized in February 1855 under the name of the Farmers' High School. Subsequently, the school officially changed its name to the Agricultural College of Pennsylvania, in order to "convince the Pennsylvania legislature to designate the school as the beneficiary of the Commonwealth's land-grant funds." This new Agricultural College ardently competed with five other schools for the Pennsylvania land-grant endowment, which it finally received in the spring of 1863.

Despite attempts by state universities, polytechnic schools and other institutions to obtain or split the land-grant endowment, acceptance of the Morrill Act provisions nullified much of the debate between rivals and those sympathetic to the agricultural colleges. With acceptance of the Morrill Act provisions, state legislatures had to act to support the new college or risk losing the land-grant endowment. In most cases, the final debate between state university supporters and agricultural college proponents took place in the state legislatures in the days leading up to acceptance of the Morrill Act provisions.[5]

As a tangible institution, land-grant colleges acted to meet the requirements of the Morrill Act by providing for education in the agricultural and mechanical arts. What followed, however, was an internal debate common to many land-grant colleges. As Earl Ross documented in *Democracy's College: The Land-Grant Movement in the Formative Stage*, from the mid-1860s through the 1880s many land-grant institutions found the goals of the Morrill Act were not being fulfilled. Often the agricultural programs did not meet the needs of the farmers, and in other instances, the mechanical arts existed in name only. Eventually, land-grant colleges would succeed in establishing advanced agricultural programs and state-of-the-art mechanic arts or engineering divisions. However, creation of such strong programs often came only after extended struggles and disputes.

One major point of contention at many institutions centered around questions about the proper relationship between agriculture and the mechanic arts. Many nineteenth-century observers viewed the mechanical arts as necessarily subordinate to agricultural pursuits. For example, when asked by the Department of Education in 1892 for statistics regarding enrollment in the agricultural and mechanic arts programs, Kansas State Agricultural College President George Fairchild responded that the Education Department wrongly "*assumed* such an entire separation of these departments from each other." Fairchild, like many of his colleagues, assumed that studies of the mechanical arts and the "application of the laws of mechanics" at land-grant institutions should include only those areas "practicable to agriculture."[6]

Indeed, land-grant institutions, proud of the new agricultural experiment stations established under the Hatch Act of 1887, boasted of this agricultural focus by emphasizing it up-front in their names (as seen with the Kansas State Agricultural College example). Assumptions about the subordinate role of mechanic arts to agriculture appeared throughout the late 1880s in records of the Association of American Agricultural Colleges and Experiment Stations (hereafter the Association).[7]

The Association had been formally organized in 1887, at the urging of Norman J. Colman, Commissioner of Agriculture with the United States Department of Agriculture (USDA). Colman created the Association as a way, he hoped, to unite the USDA with the agricultural colleges. In particular, he aimed to make the agricultural colleges and the newly established agricultural experiment stations serve as the experimental grounds of the USDA. The roots of the Association, therefore, rested firmly in the soil of America's national agricultural interests. In turn, the mechanic arts were viewed as secondary, actually subservient to agriculture. According to Commissioner Colman's 1887 opening remarks, "agriculture is the nursing mother of all our industries and is entitled to such recognition [from] . . . her handmaids—manufactures and commerce."[8]

Almost immediately, it became evident that not everyone felt comfortable giving agricultural education such an overwhelming priority. In an 1887 paper, "Industrial Education," Michigan State Agricultural College president Edwin Willits pointed out that almost every college and university had "yielded to the peremptory demand that the sciences shall be recognized in the college curriculum." Arguing that the "old school" had passed, Willits urged his fellow delegates to embrace "the age of applied science" and "to establish and foster in the United States schools of technology for instruction in industrial arts." Although Willits's presentation proved to be "full of thought and suggestion," delegates moved to delay further discussion of the ideas presented. Despite objections, George W. Atherton, president of the Pennsylvania State College, agreed that the "want of time" did not allow further discussion. The partisan Association continually drowned out engineering supporters. Other actions by the Association reflected its persistent agricultural preference.[9]

The *Proceedings* of the second convention in 1888 further illustrated the bias of the Association toward an agricultural emphasis. The mechanical arts received little consideration. In his opening remarks, speaking for the city of Knoxville and the University of Tennessee, Judge Temple lauded the benefits the Morrill and Hatch Acts had on agriculture. Temple also noted his optimism that "ample provision" would be made for the mechanic arts in the "near future." Despite his optimism, the mechanic arts garnered no further attention at the 1888 or 1889 conventions.[10]

In 1890, the federal government passed new legislation, Senate Bill 3714, better known as the Second Morrill Act, which provided "more complete endowment and support" for the land-grant colleges. The Association's Standing Committee on College Work proceeded to analyze this new law. In a report presented to the group at its next annual meeting, the committee offered recommendations on how the extra funds designated for disbursement should best be applied. The resolution stipulated

> "distinct account[ing]" for the funds spent, in the order of importance and preference: (a) for instruction in agriculture and mechanic arts; (b) for facilities for such instruction; (c) for instruction in the other branches of learning specified by the law; (d) for facilities for this latter class instruction.[11]

The Association resolution demonstrated the continued emphasis on the "importance and preference" of agricultural pursuits. Equally evident, the Association still considered the mechanical arts secondary to agriculture. Despite the subordinate position given the mechanical arts, brief praise for "civil, mechanical, and electrical engineering" programs were offered in the presidential address by James Smart of Indiana's Purdue University. President Smart's other remarks indicated his belief that technical instruction and engineering programs at the land-grant institutions were proceeding at an "equally satisfactory" pace. Smart's perception, however, did not hold true for all Association members, and mechanical arts supporters would make their position clear at the Association's 1892 meeting.[12]

By the early 1890s, mechanical arts supporters within the land-grant colleges began appealing to the Association's executive committee to win proper recognition. Mechanic arts supporters wanted creation of their own section within the Association, like those provided for agriculture and horticulture. More than that, they also wanted to change the name of the Association to reflect the mechanical arts as the second, and not subservient, agency of the land-grant colleges, as stipulated by the Morrill Act. The Association, however, perceived mechanic arts supporters' efforts as a threat to the organization. A letter to Association president George Atherton expressed the opinion that it would prove "harmful to our Association interests to have minor organizations springing up and making appeals for support." The topic subsequently appeared in the 1892 *Proceedings*, when opponents of a mechanic arts section accused mechanic arts professors of having "endeavored to start an organization in a measure to rival" the Association—the Society for the Promotion of Engineering Education (SPEE).[13]

Discussion of the Association constitution and proposed amendments had regularly appeared in the *Proceedings* each year, corresponding with the group's annual meeting. The minutes of the 1892 gathering, however, demonstrated that these discussions were not always congenial or collegial. Responding to the question regarding amendments to the organization of Association sections, James K. Patterson, president of the State College of Kentucky, made a motion to establish a section on mechanic arts. Patterson pointed out that "by the terms of the law," the mechanical arts should receive adequate acknowledgment. Patterson further argued that given the persistent emphasis on agriculture, the mechanical arts had "never found recognition in the Association." The debate that followed brought out all the areas of contention that existed between agricultural and mechanical arts supporters.[14]

Opponents of the mechanical arts argued that professors of the mechanic arts, instead of joining, had "endeavored to start an organization in a measure to rival" the Association. Supporters fired back that the Association had to do more than just "invite them" to join the Association. Additionally, they pointed out again that the name of the Association was a misnomer, granting "agriculture an improper prominence." They charged that in not recognizing the mechanical arts, the Association constitution failed to represent two-thirds of the member institutions properly. One supporter of creating a mechanic arts section suggested that such a move would be important to attract the college engineers to the Association, who previously had "felt that they were not wanted" there. Again, a motion was made to change the organizational name. Feeling overwhelmed, opponents moved to adjourn, interrupting the mechanical arts supporters. Their effort proved futile, however, and the motion to establish a section on mechanical arts carried. It remained to be seen, however, whether the engineers would support the newly created section and the Association. A second motion, calling for changing the name of the Association to officially recognize the mechanic arts, died after its referral to the executive committee for further consideration.[15]

Refusing to give up, mechanical-arts supporters at the Association's next annual meeting, in 1893, again proposed a resolution to change the name of the Association. Referred to the executive committee, the resolution to recognize the mechanical arts again died there. Despite this lack of recognition, supporters of the mechanic arts section made up a substantial constituency at the 1893 gathering. At the first official meeting of their own section, members read four papers; another three were read by title and referred to the executive committee for publication.[16]

One paper read before the section meeting was that by Iowa State professor George Bissell (unable to attend in person). Bissell's paper, "Shopwork Instruction at Iowa State College of Agriculture and Mechanic Arts, Ames Iowa," outlined the systematic procedures used in the college shop. Bissell described shopwork instruction, experimental apparatus, and the relation of the program to the departments of the college. Showing dedication to his mentor, Bissell quoted Dr. Robert H. Thurston to justify the importance of applied science. Advocating "training by systematic performance of carefully planned exercises," Thurston's Cornell model had stressed the importance of giving students firsthand experience with the applied aspects of engineering. In turn, Bissell's paper advocated fieldwork and hands on experience, serving as a model for other college engineering departments. With a successful agricultural experiment station serving as an example, Bissell argued that "fieldwork" and "hands on experience could be successfully applied to engineering.[17]

It is noteworthy that, whether writing of "applied engineering" or of "manual training and the apprentice system," the mechanic arts section presentations of 1893 all related the mechanic arts to agricultural applications. To reassure the Association of its interest in maintaining a working relationship with the larger group, the section in the following year established a committee to specifically "define the scope of the work of the Section on Mechanic Arts."

The resolution subsequently adopted by this committee demonstrated the ongoing disparity between the agricultural and mechanical arts, as well as the inability of the mechanic arts section to adequately define itself and its relation to agricultural education. The committee resolved to solicit "papers dealing mainly with those subjects which fall under the designation 'Mechanic Arts' in their special relation to the work of colleges of agriculture and mechanic arts." In essence, the committee resolution failed to clarify either the object of the section or the section's relation to the larger Association. Recognizing this shortcoming, section committee members met several times throughout the following year in attempts to develop and further define the "scope" of the mechanic arts section. Though consulted, the Association executive committee offered no guidance or support to the section.[18] Following those deliberations by the section committee on mechanic arts, Secretary Anderson came before the general Association to report. Anderson noted the effort had been one-sided, examining strictly the "correlation of mechanic arts and agriculture." Anderson noted at the outset that the recently established Society for the Promotion of Engineering Education covered the broader field of engineering, meaning that by comparison, the Association section's "range of work" would be "a limited one." Anderson's section report maintained that "by welldirected efforts," the mechanic arts could provide a valuable complement to the agricultural side of the Association. The report further suggested that if the Association truly desired representation from "the mechanical side" of the institutions, "the agriculturalists must take the initiative" to attract such delegates. Anderson noted that most "men interested in mechanic arts [were] indifferent toward doing work for the Association claiming that the Association [was] essentially agricultural and it [was] the intention of the agriculturalists to keep it within such limits."[19]

Resentment over the perceived lack of recognition for the mechanical arts again generated an outbreak among Association members. As Anderson neared completion of the section report, he was interrupted by a motion to adjourn. What followed, according to section minutes, was a "discussion [that] brought out very forcibly that the Section on Mechanic Arts [was] not supported by the general Association, and in order to induce prominent educators in mechanical lines to take an active interest in the work of the section, some recognition must be given to mechanic arts in the name of the Association inasmuch as agriculture [was] specifically recognized." As the *Proceedings* for the 1895 meeting indicated, the section on mechanic arts not only lacked recognition by the Association, it also failed to gain the respect of most of its members.[20]

Pressing ahead with the quest to gain proper recognition by the larger association, the mechanic arts supporters rallied the following year to demonstrate "that the work in the mechanic arts side of the agricultural colleges" had been extended and strengthened over recent years. To move the section forward even more, mechanic arts supporters attempted to persuade the larger Association to assist them in their efforts. The section report for 1896 voiced support for two bills that had been presented in Congress that year, known as the "State Engineering and Experiment Stations Bill" and the "Wilson-Squire Engineering and Education Bill."[21] In section minutes, supporters described these measures as extremely important to the future of the mechanic arts in America.[22]

In delivering the section report, J. W. Lawrence of Colorado outlined a survey he conducted which asked land-grant institutions to comment on "what advances, if any, had been made in the studies and work of the departments operating along the line of the mechanic arts" and "the general condition of the work." In this regard, Lawrence requested the same information from the engineering departments that the Association recommended the agricultural experiment stations provide in their annual report. Lawrence's survey results indicated "an evident desire in the mechanical departments of the colleges, and more especially in the younger ones, to raise the standard of scholarship." The section report suggested that the pending legislation would do much to improve this situation regarding "facilities for instruction . . . advanced laboratory work . . . [and] material equipment." The survey, according to Lawrence, indicated "hearty endorsements" by college presidents and department professors for the State Engineering Experiment Stations Bill. As a final point, the section report indicated that the pending legislation was "similar to the Hatch Act" and "the logical sequence of previous legislation." The secretary concluded the report soliciting the support of the Association. The chairman accepted the section report without objection, but criticism of the station idea remained to be heard.[23]

By 1896, the mechanic arts section had generated much interest and support throughout the land-grant colleges. Still, when the section again asked the executive committee to recognize the mechanic arts in its name, what followed amounted to an identity crisis within the Association itself. Organized in 1887 to benefit agriculture and the USDA, the Association of American Agricultural Colleges and Experiment Stations had, by 1896, evolved into a land-grant college governing body offering guidance on curriculum and professional development. The Association's views on engineering education remained disjointed and secondary. Thus, when asked to revise and include mechanic arts in the name of the Association, executive committee members even considered abolishing all the sections and restructuring the Association as a "conference for executive officers of Colleges and [Experiment] Stations." There "developed a wide difference of opinion," and even this last "radical proposition" received consideration.[24]

As the Association struggled with its own identity, mechanic arts supporters found little guidance or assistance for promoting their interests. Thus, in 1896, when the Hale-Dayton bill to establish engineering experiment stations was proposed in Congress, Association leadership offered little support for the bill. Despite this lack of backing from their elected leaders, some individual members discussed and openly supported the legislation. Most notably, Cornell University president Robert H. Thurston addressed the 1896 Association convention to encourage mechanical engineering and to urge support for the pending legislation to establish experiment stations. Thurston presented his talk as a consideration of the "professional education of the mechanical engineer," addressing the Association's constitutional provision on curriculum and professional development. Thurston's paper addressed controversial issues, asking "What are the precise purposes of such education and training" that the agricultural colleges provide mechanical engineers? and "How can the best contemporary practice be improved?" He concluded that engineers urgently needed advanced courses, scientific investigation and the study of laboratory methods, for which purpose he openly promoted the pending federal legislation. Despite discussion of Thurston's remarks and a presentation titled, "Engineering Experiment Stations," in the end, opponents argued that it was for the recently established Society for the Promotion of Engineering Education (SPEE) to solicit support for the engineering experiment station bill.

As it turned out, SPEE members did not fully endorse the legislation either. Organized in 1893, the Society for the Promotion of Engineering Education had grown out of the Chicago meeting of the World's Engineering Congress. As a "meeting of teachers of engineering," the SPEE had few initial goals other than to serve "teachers." Speaking before the SPEE in August 1896, William Aldrich of West Virginia University outlined the arguments for and against the proposed

Hale-Dayton Bill. Modeled after the Hatch Act, which had endowed the agricultural experiment stations, the engineering station bill met with a variety of dissent. Some critics argued that engineering was not an experimental science, while others maintained that engineering experimentation was not local and that results could be reapplied broadly. Still others insisted that the federal government should not support state manufacturing and industry, while yet another group urged endowment of one good laboratory and not fifty copies. Additional arguments against the bill centered on the agency with oversight—the U.S. Navy. Critics warned of the dangers of maintaining too close a government connection, pointing to the intimate, but sometimes adverse, relationship between the agricultural stations and the U.S. Department of Agriculture. In response, advocates of engineering stations shrugged that warning aside, noting that the Navy would merely serve as a repository of station journals. Unable to reach a consensus, SPEE members remained divided in their support of the legislation.[25]

Although the bill received public support, the petitions and correspondence sent to members of Congress were not coordinated. An examination of the *Journal of the Senate* showed that all of the letters in favor of the legislation were arbitrarily distributed among the Committee on Public Lands, the Committee on Education and Labor, the Agriculture and Forestry Committee, and the committee actually responsible for the bill, the Naval Affairs Committee. This disorganization undoubtedly undercut any sense of popular backing for the legislation. Lack of agreement among engineers further hampered the legislation.[26]

The failure of the Association of American Agricultural Colleges and Experiment Stations to promote the Hale-Dayton Bill led some members to question that organization and the quality of its leadership. In particular, Charles S. Murkland, president of the New Hampshire College of Agriculture and Mechanic Arts, addressed the August 1897 meeting of the SPEE as a member of both the SPEE and the Association. Murkland's presentation identified the problem as "The Agricultural College in Its Relation to Engineering Education." Tracing the root of the problem to the name "Agricultural College," Murkland protested that such a name proved "a misnomer" and generally "misleading." Beyond these complaints about the improper name, Murkland's primary objection to those institutions' current state centered around the lack of "uniformity as to the kind or amount of instruction immediately pertaining to engineering. The differences concern . . . entrance requirements, courses of study, and degrees." Although the Association had appointed a committee in 1895 "to study entrance requirements for engineering programs," Murkland told the SPEE audience that the "standard" accepted "was not in any sense ideal." He advised the SPEE that if it were to decide on a "standard of preparation," he believed that "any suggestion of this kind . . . would be welcomed by the Association of American Agricultural Colleges and Experiment Stations, and would be put into practice as widely as possible." Murkland felt frustrated that the Association's present leadership continually failed to consider the needs of engineering education and thus had failed to support federal legislation addressed to that end.[27]

Less than two years later, the federal government once again considered legislation to establish engineering experiment stations. During the Fifty-sixth Congress, several bills proposing such a measure were introduced in both the House of Representatives and the Senate. The original bill, HR 982, called for "endowment and support of the mining schools in the several States and Territories." HR 982 was fashioned after the Hatch Act, based on the mistaken premise that mining experiment stations deserved the same type of support "as already provided for the agriculture and mechanical arts." In fact, the mechanical arts had never been equally provided for in land-grant history. A second bill, HR 7725, called for "mining experiment stations." As ultimately revised, the bill S 3982 called for "endowment, support, and maintenance of schools or departments of mining and metallurgy" at land-grant institutions.[28]

The Congressional *Docket* and committee records indicate that the legislation soon received some favorable recommendations. Supporters included former Association president George Atherton of Pennsylvania State College and other land-grant presidents such as E. Benjamin Andrews of Nebraska. Still, correspondence from the presidents of state institutions, such as the University of Idaho, objected to the idea of the federal government creating yet "another" land-grant college endowment. The jealousies among different institutions in the same state continued, as advocates of land-grant college interests and supporters of state universities jockeyed for support. Responding to the proposal to establish schools of mining at the land-grant colleges, the Board of Trustees of the University of Alabama petitioned "Congress to divert the fund" to the University, rather than "its rightful place, the Alabama Polytechnic Institute." In turn, the land-grant institution president and alumni petitioned members of Congress to remedy that "gross act of injustice" initiated by the University of Alabama.[29]

Despite the fact that S 3982 provided federal support for mining and metallurgical engineering, a special survey conducted by the Association suggested that some members still felt an urgent need for federal endowment to establish more general "engineering experiment stations." The Association's executive committee, however, decided that "the time had not yet come for urging before Congress passage of such legislation." And although Pennsylvania State College president George Atherton favored the legislation and actually appeared before the House Committee on Mines, Association president and director of the Pennsylvania State Agricultural Experiment Station, Henry Prentiss Armsby did not. The internal conflict at Penn State was not uncommon at other land-grant institutions. Ultimately, the bill failed to garner enough votes in the House of Representatives. Once again, engineering supporters felt betrayed by the land-grant governing body and legislators.[30]

This failure to support the legislation brought agriculture and mechanical arts supporters into conflict again. Upset with the Association for not generating support for the measure, mechanical arts supporters voiced their discontent with the executive committee and President Armsby in their annual report. Engineers still resented the feeling of being treated as second-class citizens inside the Association and within their own institutions. At Pennsylvania State College during Armsby's tenure as a professor, the dean of the College of Agriculture, and also the first and only director of Penn State's agricultural experiment station (1887–1907), agricultural interests continued to dominate the institution.

In the 1900 report from the Association Committee on Engineering Experiment Stations, Secretary Charles Murkland of New Hampshire noted that the committee stood "ready to respond to any call" from the executive committee. Since they had "not been called upon" by the executive officers, Murkland believed that "may indicate the status of the project it stands for." The secretary also informed the executive committee that the opportunity to support the establishment of engineering experiment stations had again passed. Murkland concluded his report by noting that no reason existed to continue the Committee on Engineering Experiment Stations. Again, Association leadership missed a chance to voice explicit support for the mechanic arts.[31]

The following year, Harry W. Tyler of M.I.T. delivered the annual report for the section on mechanic arts. Straying from the usual course, Secretary Tyler informed the executive committee he was going to "take the liberty to speak . . . particularly of the scope and function of the mechanic arts section believing that the Association [had] still to reach a better solution of this problem." In the course of his report, Tyler warned that the mechanic arts section should not expect or take for granted "that professors of engineering will or should be attracted to our annual meetings." Mentioning the recent meeting of the SPEE in Buffalo, Tyler noted the duplication of work conducted by both groups and observed that the SPEE was ahead of the Association in promoting mechanical arts and especially curriculum development. Tyler concluded his section report by noting possible changes the executive committee could consider making to remedy the problem. With the SPEE attracting the land-grant college engineers, Tyler specifically wanted the Association's executive committee to determine the scope of the mechanic arts section. The Association's executive committee accepted Tyler's report, but apparently ignored those suggestions and questions, giving the report no further consideration. Perhaps the committee had accepted the inevitability of its loss of the engineering constituency. On yet another occasion, the Association had failed the mechanical arts supporters.[32]

Though Association members who were engineers openly expressed their anxiety, the section continued to meet over subsequent years, drawing fewer attendees and supporters each time. Finally, in 1906, after another failed attempt to gain recognition and support from the larger Association, the mechanic arts section dissolved itself. In one last bid for support, Iowa State College president A. B. Storms reminded the Association of the original Morrill Act provisions, speaking of the important and separate roles of agriculture and mechanic arts. Storms advised the group to recognize and support the mechanic arts, just as it had done for agriculture years earlier. Pointing to the benefits of experimentation as seen in the agricultural stations, President Storms called on the Association to support establishment of strong experiment stations for the mechanical industries.[33]

To support his case, Storms cited the achievements at his own Iowa State College. Back in the early 1890s, Robert Thurston's former Cornell students Anson Marston and George Bissell had established the mechanical laboratory at Iowa State College shortly after their arrival in Ames. To further support the engineering profession, Marston crafted the engineering curriculum to emphasize more scientific research techniques and basics in mathematics and physics. The Engineering Department also conducted numerous tests and investigations for Iowa manufacturers. These engineers had felt frustrated for years by stalled efforts to gain federal support for their work. Encouraged by the "constant demands for assistance from the various mechanical industries of Iowa," these advocates at Iowa State College instead initiated efforts in 1903 to gain state support for an engineering experiment station. By April 1904, Iowa State College had the nation's first state-sponsored engineering experiment station. Thus, when President Storms spoke in 1906 of the benefits offered by engineering experiment stations to the discipline and to the public, he spoke from firsthand experience.[34] He felt confident of the benefits of experimental engineering research, and thus urged his fellow Association members and the group leadership to finally support this cause.[35]

The following year, Senator McKinley of Illinois sponsored another federal "Bill To establish engineering experiment stations at land-grant colleges." A bill with the same title, sponsored in 1909, suffered similar defeat. By that point, the Association's section on mechanic arts had been dissolved; thus, the Association felt little to no pressure to voice support for McKinley's bill, and indeed, it made no effort to do so.[36]

Again in 1916, the Association failed to support renewed calls for establishing federally funded engineering stations. In November 1916, Iowa State College dean of engineering Anson Marston addressed the Association on the topic of "Proposed Legislation to Establish Engineering Experiment Stations." He opened by reminding listeners that the original intent of the Land Grant College Act of 1862, Section 5, called for "systematic research at each land-grant institution, and for the publication of the results." Marston noted that the research side had been "much neglected . . . prior to the passage of the Hatch Act." Praising agricultural research and the distribution of that information, Marston emphasized the usefulness of systematic research. Specifically, Marston praised the intent of the Hatch Act to establish stations where systematic or applied research could be conducted. Furthermore, Marston informed Association members that "logically, Congress should have provided at the time of the passage of the Hatch Act for engineering experiment stations as well as agricultural stations and doubtless this action might have been taken at that time if engineering science had been as backward as agriculture. It was the crying needs of agriculture, rather than any intentional neglect of mechanic arts, which led to the omission of provision for engineering research."[37]

Marston again cited the Morrill Act, noting that the need for engineering experiment stations had become "increasingly apparent, especially at the land-grant institutions, which have by federal law a most intimate relation to our industries and to the industrial classes." In describing past attempts to federally establish engineering experiment stations, Marston elaborated on the most recent example and the reasons for its failure. He noted that the Newlands Bill of 1916 had failed to pass "partly because of lack of active effort by institutions affected and partly because of certain jealousies and conflicts of interest." Marston hoped that it would yet be possible for those "jealousies and strife" to subside, saying that it would be "essential that the Executive Committee of the Association of American Agricultural Colleges and Experiment Stations assume the principal direction of the movement." In truth, the core of these jealousies and conflicts of interest dated back to the original Land-Grant Act of 1862. Points of contention between supporters of agriculture and those who supported the mechanic arts would not vanish so easily. Again in 1918, the Association declined to support proposed engineering station legislation, and again the bill failed. Clearly, the Association was not yet ready to alter its emphasis on promoting agricultural interests above all.[38]

As more and more Americans came to recognize the benefits of mechanical arts and experimentation, however, a number of states chose to follow Iowa's lead in sponsoring their own engineering experiment stations. By 1916, a dozen states had established stations. Two dozen more would be established within fifteen years. By 1920, the Association of American Agricultural Colleges and Experiment Stations would finally come to acknowledge the important role of the mechanic arts. Indeed, the irony of the situation is that by 1920, the Association began appealing to mechanic arts supporters, engineering colleges and state-sponsored engineering experiment stations for their support. The important role of engineering education and research had repeatedly proved its merit. The Association recognized that engineering programs at land-grant colleges competed with similar programs at state universities and private colleges. Therefore, the Association decided it would be in its own best interest to support those programs based at the former *agricultural* colleges. Also, members realized that agricultural engineering needed the mechanic arts and the cooperation of forward-thinking engineers in order to thrive. In 1920, the Association of American Agricultural Colleges and Experiment Stations formally recognized the mechanic arts as an equal and official section. To further show its support, the group officially changed its name to the Association of American Agricultural and Mechanic Arts Colleges and Experiment Stations.

Such measures, however, had come too late for the battle over federel funding of engineering stations. The Association's lack of support for the mechanic arts had directly contributed to the repeated failure of proposed legislation for that purpose. The land-grant colleges had started out in conflict with state university and other institutional proponents. Once established, agricultural colleges almost immediately faced the problem, within their own institutions, of strife and jealousies between the agricultural and mechanic arts branches. Their failure to gain recognition and support from their own governing body further frustrated mechanic arts supporters and ultimately led them to turn to the SPEE for guidance and support. As such, land-grant college engineering departments found themselves at odds with agricultural program supporters. This internal strife not only weakened the land-grant colleges' position in urging federal legislation, it also provided ammunition for state university proponents who stressed the weaknesses of the agricultural college engineering programs. Tired of waiting for federal endowment, those institutions that valued a more scientific approach to engineering solicited their respective state legislatures to recognize and support engineering experiment stations. In understanding the roots of this frustration, one can better understand the history and role of land-grant college engineering experiment stations.

Notes

1. Specifics on the origin and intents of the Morrill Act as found in Earle D. Ross, *Democracy's College, The Land-Grant Movement in the Formative Stage* (Ames, Iowa State College Press, 1942); Ross, *A History of The Iowa State College of Agriculture and Mechanic Arts* (Ames, Iowa State College Press, 1942); and Alan I Marcus, *Agricultural Science and the Quest for Legitimacy: Farmers, Agricultural Colleges, and Experiment Stations, 1870–1890* (Ames, Iowa State University Press, 1985); Quotes as cited in the original legislation, The Land Grant College Act of 1862, commonly referred to as the Morrill Act, specifically, "An Act Donating Public Lands to the Several States and Territories Which May Provide Colleges for the Benefit of Agriculture and Mechanic Arts," *United States Statutes at Large*, 1862, 12:503–5.

2. Numerous authors have written of the professionalization of engineering and the many elements it entailed—engineering education, societies, journals, research, etc. See, for example, Alan I Marcus and Howard Segal, *Technology in America: A Brief History* (Orlando, Harcourt Brace Jovanovich, Publishers, 1989), 165–73, 235–36; Bruce Sinclair, *A Centennial History of the American Society of Mechanical Engineers, 1880–1980* (Toronto, University of Toronto Press, 1980); Edwin T. Layton, "Mirror-Image Twins: The Communities of Science and Technology in 19th-Century America," *Technology and Culture* (October 1971), 562–80; Peter Meiksins, "The Revolt of the Engineers Reconsidered," *Technology and Culture* (April 1988), 219; Bruce Seely, "Research Engineering and Science in American Engineering Colleges, 1900–1960," *Tech-*

nology and Culture (April 1993), 344; On the World War I account, see Daniel Kevles, "Federal Legislation for Engineering Experiment Stations: The Episode of World War I," *Technology and Culture* (April 1971), 182.

3. Layton, "Mirror Image Twins: The Communities of Science and Technology in 19th-Century America."

4. The Iowa legislature directed the board of trustees to establish the engineering experiment station under Iowa Code, Chapter 156, Section 1, dated April 13, 1904; unlike the other early station, the Illinois station was established by the board of trustees in late 1903, without initial state support.

5. Ross, *A History of Iowa State*, and *An Historical Sketch of the Iowa State College of Agriculture and Mechanic Arts* (Ames, published for the Semi-Centennial Celebration, June 1920); The Pennsylvania State University example as found in Michael Bezilla, *Engineering Education at Penn State, A Century in the Land-Grant Tradition* (University Park, The Pennsylvania State University Press, 1981), 3–5. See also Robert C. McMath, Jr., et al., *Engineering the New South, Georgia Tech, 1885–1985* (Athens, University of Georgia Press, 1985).

6. Letter to President George W. Atherton, from Kansas State Agricultural College president George T. Fairchild, dated November 29, 1892, The Penn State Room, Pennsylvania State University, Pattee Library, University Park, Pa.; George W. Atherton, Box 5, File Z, "Land Grant Colleges: Statistics, 1892–96." Commissioner Colman's remarks found in *Proceedings of a Convention of Delegates From Agricultural Colleges and Experiment Stations*, U.S.D.A. Miscellaneous Special Report No. 9 (Washington D.C., G.P.O., 1887).

7. Agricultural experiment stations were officially established under provisions of the 1887 Hatch Act. For specifics on the original legislation and background, bill passage and outcome, see Marcus, *Agricultural Science and the Quest for Legitimacy*, especially 127–221.

8. Marcus, *Agricultural Science and the Quest for Legitimacy*, 195, 212–13; *Proceedings of a Convention of Delegates from Agricultural Colleges and Experiment Stations*, U.S.D.A. Miscellaneous Special Report No. 9 (Washington, G.P.O., 1887).

9. *Proceedings*, 1885, 55, 57, 61, 65–66.

10. Ibid., 1889, 1890. A review of the *Proceedings* contents shows an emphasis on agricultural education and experiment station details. Discussions explored such topics as "How can the stations reach and interest the farmers?" "How can the Department of Agriculture assist the stations?" "How can stations cooperate?" "Does the Hatch act need amendment?" and "Station organization and methods." *Proceedings*, 1889, 16.

11. *Proceedings*, 1891, 20, 139.

12. Ibid., 41.

13. *Proceedings*, 1892, 74–76; Letter to George W. Atherton from Henry E. Alvord, President, Maryland Agricultural College, dated April 20, 1892, The Penn State Room, Atherton, Box 5, File G, "A.A.A.C.E.S. & Experiment Stations, 1891–

93"; *Proceedings*, 1894, Executive Committee Report, 64. The SPEE officially organized in 1893.

14. *Proceedings*, 1892, 74.

15. *Proceedings*, 1892, 74–76; Letter to George W. Atherton from Henry E. Alvord, president, Maryland Agricultural College, dated April 20, 1892, The Penn State Room, Atherton, Box 5, File G, "A.A.A.C.E.S. & Experiment Stations, 1891–93;" *Proceedings*, 1894, Executive Committee Report, 64. Though not specifically stated, the Association saw as its rival the Society for the Promotion of Engineering Education.

16. *Proceedings*, 1894, 64. The 1892 *Proceedings* actually show that more mechanical arts supporters arrived at the convention prepared with papers to deliver than did the other four sections.

17. *Proceedings*, 1894, 65–67; Gilkey, *Anson Marston*, 12–14.

18. *Proceedings*, 1894, 89.

19. *Proceedings*, 1896, 18–19.

20. Ibid., 97.

21. An examination of the Congressional Record indicates that four bills to establish engineering experiment stations made it to the House floor. HR 3618, introduced January 10, 1896, was referred to committee. HR 5836, introduced February 11, HR 6452, introduced February 24, and S 2301, introduced February 27, all died in their respective committees. Without access to the committee records, the nature of the debate surrounding these bills remains elusive. Regardless, these bills did generate interest and debate outside of the Congressional committees.

22. *Proceedings*, 1897, 26–28, 111–18.

23. Ibid., 27–28.

24. "Confidential" letter to George Atherton, from Henry E. Alvord, secretary, A.A.A.C.E.S. Executive Committee, dated June 15, 1897, The Penn State Room, Atherton, Box 5, File H, "A.A.A.C.E.S., 1894–99."

25. *Society for the Promotion of Engineering Education, Index to Volumes I–XX, 1893–1912 of the Proceedings*, "Origin of the Society," 3; William S. Aldrich, "The Hale Engineering Experiment Station Bill," *Proceedings of the Society for the Promotion of Engineering Education* 4 (1896), 187, and discussion thereafter, 192–215. On relations between the USDA and the agricultural experiment stations, see Marcus, *Agricultural Science and the Quest for Legitimacy*.

26. Letters, petitions and resolutions in support of the Hale-Dayton Bill as found in the Journal of the Senate, *Congressional Record*, 1st Session, 54th Congress (1896), 185, 199, 208, 233, 331.

27. "The Agricultural College in Its Relation to Engineering Education," *Proceedings of the Society for the Promotion of Engineering Education*, 1897, 295–96, 301–3.

28. HR 982, "A Bill To apply a portion of the proceeds of the public lands to the endowment and support of mining schools in the several States and Territories,

for the purpose of extending similar aid in the development of the mining industries of the nation as already provided for the agricultural and mechanical arts," introduced in the House, December 5, 1899; HR 7725, "A Bill To establish mining experiment stations, to aid in the development of the mineral resources of the United States, and for other purposes," introduced in the House, January 31, 1900; S 3982, "An Act To apply a portion of the proceeds of the sale of the public lands to the endowment, support, and maintenance of schools or departments of mining and metallurgy in the several States and Territories in connection with the colleges for the benefit of agriculture and the mechanic arts established in accordance with the provisions of an Act of Congress approved July second, eighteen hundred and sixty-two," dated May 16, 1900. All three as found in the U.S. National Archives and Records Administration, HR 56A-F25.1, "Committee on Mines and Mining, Various Subjects."

29. Letter from Chancellor E. Andrews, University of Nebraska, to Congressman David H. Mercer, chairman, Committee on Public Buildings and Grounds, dated November 22, 1900; Letter from Chairman Mercer to R. O. Crump, chairman, Committee on Mines and Mining, dated December 5, 1900; Letter from J. P. Blanton, president, University of Idaho, to Senator George L. Shoup, dated March 5, 1900; Letter from Senator Shoup to Senator William M. Stewart, chairman, Senate Committee on Mines and Mining, dated March 17, 1900; Letter from "C. I.," president, Alabama Polytechnic Institute, addressed "Dear Brother," dated January 29, 1901. All materials as found accompanying copies of original legislation (see note 13), U.S. National Archives and Records Administration, Sen. 56A-F22, Mines and Mining, S. 3109 – S. 5589.

30. Atherton citation as found in the Congressional *Docket*, Committee on Mines and Mining, 56th Congress, meeting dated May 21, 1900, U.S. National Archives and Records Administration (U.S. N.A.R.A.), *Docket*. See also Committee report, H.R. Report No. 1631, dated May 21, 1900; Notes on Henry Prentiss Armsby, biographical file, The Penn State Room. Biographical reprint, "Henry Prentiss Armsby," *The Journal of Nutrition* 54, no. 1 (September 1954).

31. "Report of Committee on Engineering Experiment Stations," *Proceedings of the Association of American Agricultural Colleges and Experiment Stations*, 1900, 40–41.

32. *Proceedings*, 1901, 33–40.

33. Ibid., 1907, 97–98.

34. Anson Marston papers, 11/1/11, box 9, Archives & Special Collections, Parks Library, Iowa State University. The Iowa legislature officially recognized the station in April 1904 appropriation legislation. Though organized in December 1903, the University of Illinois station was established by the board of trustees, not the state legislature.

35. "An Engineering Experiment Station," *The Iowa Engineer* 3, no. 1 (June 1903), 71; Letter from Dean Anson Marston to President A. B. Storms, dated October 6, 1903, Iowa State University Archives, Marston (11/2/1, box 7-2).

36. H.R. 9230, "A Bill To establish engineering experiment stations at land-grant colleges," 60th Congress, 1st Session, U.S. N.A.R.A. H.R. 60A-F2.1-F2.6, "Miscellaneous"; S. 2768, "A Bill To establish engineering experiment stations at land-grant colleges," 61st Congress, 1st Session, U.S. N.A.R.A., Sen.61A.B1, "S. 2736 – S. 2770."

37. *Proceedings*, 1917, 26. For additional comments regarding engineering education and the profession by Anson Marston, see Herbert J. Gilkey, *Anson Marston: Iowa State University's First Dean of Engineering* (Ames, Iowa State University Press, 1968).

38. *Proceedings*, 1917, 26–28.

■ Creation of the Modern Land-Grant University

Chemical Engineering, Agricultural By-Products and the Reconceptualization of Iowa State College, 1920–1940

Alan I Marcus and Erik Lokensgard

By the 1910s, the engineers of Iowa State College had emerged from agriculture's shadow and created an independent identity for themselves. No longer an after-thought in a predominantly farming state, ISC engineers set their sights far beyond the college's students, claiming parity with the college's agricultural wing by tying themselves to Iowa's manufacturers and businessmen. Through activities that would be codified and formalized into engineering extension in 1913, the engineers catered to Iowa's manufacturers in several ways and therefore made themselves virtually indispensable to that group. ISC engineers provided technical help in building infrastructure and establishing processes as well as training persons to supervise and superintend the mechanic arts in the state's public schools. This latter effort would provide the next generation of technicians to work in the state's manufacturing sectors.

The two most prominent and well-established types of engineering—mechanical and civil—led the transition. Guided by Anson Marston, professor of civil engineering and later dean of the engineering college, civil and mechanical engineers embraced the new agenda and directed the construction of roads, designed meat packing plants and trained mechanical arts teachers to work in the public schools.[1] But while the mechanical and civil engineers dominated the engineering faculty, other engineers found no such simple solution. Chemical engineers were one such group.

Chemical engineering was a new enterprise at ISC. In 1910 Iowa State introduced a few courses under the rubric "industrial chemistry" and in 1913 it approved an undergraduate course of study resulting in a chemical engineering degree. The first degree was awarded a year later.[2] Yet at Iowa, this seemingly minor and newly created branch of engineering would come to hold a far more significant place in the history of the college than its meager start or paltry numbers might suggest. As early as the 1920s and 1930s chemical engineering became the vehicle through which to implement the revamping and reconceptualization of the college. Chemical engineering provided the basis to integrate college faculties, divisions and departments to create something very much like the modern research university.

That Iowa State lacked a formal chemical engineering program before 1913 was not particularly unusual for a land-grant school. Chemical engineering as a distinct type of engineering was of relatively recent vintage. Not until 1908 did a national group coalesce to form a professional organization, the American Institute of Chemical Engineering. Even then the discipline's raison d'être was questionable. What did a chemical engineer do that was different than a mechanical engineer? True, chemical engineers dealt with chemical processes and establishments. But were not chemical facilities simply a type of manufacturing entity and therefore little different from other manufacturing forms, which long had fit neatly in the province of mechanical engineers? Certainly the early work at Iowa State suggested that chemical engineering differed little from its mechanical counterpart. Charles A. Mann, the professor in charge of chemical engineering, organized the curriculum into five equally important manufacturing processes: chemical manufacture, electrochemistry, metallurgical chemistry, gas manufacture, and food processing. In 1915, Mann upped the stakes but did little to change the emphasis. He required undergraduates concentrating in chemical engineering to engage in research and prepare senior theses before they could receive bachelor's degrees. Two years later the college, at Mann's behest, began to grant master's degrees—explicitly research degrees—in that discipline. Both the bachelors' and masters' theses written prior to Armistice Day reflected Mann's commitment to wide-ranging mechanical engineering-like research. No two of the eleven theses—nine by undergraduates—focused on a similar problem, nor were they, as a group, confined to a single branch of chemical engineering. The theses represented all five major divisions as students considered such diverse topics as the extraction of aluminum and iron sulfates from coal wash, the tensile strength of paint films, the conversion of paper-mill waste into cement, and the corrosion of iron.[3]

While chemical engineering was a discipline in search of itself, early twentieth century chemistry was well established. In fact, chemistry (and physics) was the exciting and cutting-edge discipline of the early twentieth century. This was especially true of organic chemistry. Chemically prepared dyes, pharmaceuticals, and foodstuffs—all variations of organic chemistry molecules—promised to improve the quality of modern life. Discovery of celluloid, and then in 1907 of Bakelite, the first two plastics, held even greater potential. Bakelite particularly indicated a possible revolution in chemical manufacture. This new substance, moldable to human whims and vicissitudes, could be fashioned into virtually anything a human imagined.[4]

Yet industrial chemistry, especially industrial organic chemistry, remained tantalizingly out of reach to most American manufacturers. Outside of the armament and paint industries, Germans held virtually all the important industrial chemical patents, especially in organic chemistry, and international law forbade the American chemical industry to breech them. The German government permitted the nation's chemical companies to share patents and combine to monopolize the industry worldwide. Faced with this juggernaut of resources and means, the American chemical industry had little ability to compete in many of these exciting new areas.[5]

The situation would clarify itself considerably during World War I. The Chemical Foundation, a consortium of American chemical manufacturers and firms, urged abrogation of German patents. The Foundation argued that Germany as a combatant held precious monopolies on numerous critical processes, which unless declared null and void would severely hamper the American war effort. The federal government agreed and invoked its war powers to seize the patents, which it sold at marginal rates to well-heeled American companies. Distributing the patents among established chemical manufacturers guaranteed a quick start-up. By the early 1920s, the American chemical industry dominated the American scene.[6]

The heady atmosphere for American chemistry that greeted the end of World War I contrasted greatly with the situation facing American farmers. Cessation of hostilities produced a precipitous drop in agricultural prices and a deep farm depression. In Iowa, the nation's preeminent agricultural state, the situation seemed desperate and the college was called upon to help. Led by Raymond Allen Pearson, ISC president, the college as the state's "agricultural, scientific institution" worked to provide farmers a timely assist. Under Pearson's stewardship, agricultural history found its way into the curriculum, an agricultural economics department was formed, and rural sociology gained programmatic status. Corn-breeding work became imbued with renewed vigor, veterinary medical research continued apace, and members of the agronomy department embarked on a soil-building program.[7]

Mann's resignation from the college faculty shortly after World War I offered Pearson a special opportunity to reformulate chemical engineering, to take into account the exciting new ventures in chemistry. As Pearson deliberated over the prospects for the discipline within the ISC, the college's board of trustees took care to remind him that it was "particularly desirous of having something done for the farmer." After an extensive search, Pearson decided to entrust chemical engineering's future at ISC to Orland R. Sweeney, who accepted the offer and in 1920 took over the reins of the department.[8]

Sweeney seemed an especially promising choice. Although still a young man, he came to Iowa with a history of combining chemical engineering and agriculture. As an undergraduate at Ohio State University he had worked on the destructive distillation of corncobs. After he received his PhD, from the University of Pennsylvania in 1916, his interest in the potential of farm materials accelerated. He took a position at North Dakota State College, where he not only erected a "paint fence" for the systematic testing of linseed-oil-based paints but also investigated the possibility of substituting soybean oil for linseed. He even tried to develop a disposable baby diaper composed of peat. Although the Kotex Company expressed interest in Sweeney's creation, it chose not to make the product.[9]

Sweeney's efforts prior to his arrival in Ames suggested his orientation. Each dealt with agricultural waste materials, which would prove to be the major theme of his Iowa State work. He hoped to take "something that nobody else wants" and work "on it to make something useful." The notion that industries based on agricultural wastes would boost farm incomes loomed prominently in Sweeney's thinking. Such industries would both create new manufacturing sectors in the Iowa economy—a now well-established engineering goal—and allow farmers to turn what had been waste into profits with little additional expense. Sweeney boldly predicted that these new industries could provide farmers an extra billion dollars a year. Thus Sweeney's view of chemical engineering research at Iowa State focused almost exclusively on chemical manufacture and within that area stressed farm wastes as raw materials (Iowa lacked petrochemical resources, the raw material from which American chemical industries gleaned their organic chemistry molecules). As he expressed it in a publication aimed at students, emphasis in chemical engineering at Iowa State would be "laid upon the economic recovery of byproducts, especially the large amounts of waste material resulting in agriculture, and the industries utilizing agricultural products."[10]

Choices for senior thesis topics in chemical engineering were the most immediate manifestation of Sweeney's program. Of the thirty-two theses written during his first three years, at least twenty-six explored some industrial use for agricultural wastes. These thesis topics indicated a profound change in the college's chemical engineering department. Before Sweeney assumed control, only one graduate and one undergraduate had examined subjects that touched on agriculture.[11]

Orland R. Sweeney was determined to reorient the chemical engineering faculty. None of the department's continuing members chose to engage in agricultural waste research and all left within a few years of Sweeney's appointment. He replaced them with former graduate students who had done their theses on agricultural waste industries. After this initial housecleaning, Sweeney did not let up. During the interwar period, those faculty members who did not share Sweeney's commitment to the creation of farm-waste-based industries did not remain in Ames. Dabbling in this research was not enough; those who stayed on the periphery rather than placing farm-waste research as their top priority soon went elsewhere. Conversely, those who accepted Sweeney's vision often had exceptionally long tenures at Iowa State. For example, H. A. Webber and L. K. Arnold both signed on in the mid-1920s, diligently pursued farm-waste studies, and remained at the college for over three decades.[12]

Sweeney then made over the chemical engineering department in his own image. The work there consistently showed his influence as it developed through several phases. His crew sorted out options by first investigating, in the early 1920s, the commercial possibilities of a wide variety of agricultural wastes. Sweeney, his colleagues, and their students fermented beet sugar pulp to produce alcohol, acetic and butyric acids, and paper pulp, and recovered pectin from the extracts of apple and beet processes. They also explored plastics made from peanut, cottonseed, and oak hulls, and from straw and they tried to hydrogenate cottonseed oil. Yet these early years were the "era of the corncob" because Sweeney and his cohorts quickly decided to devote themselves to its potential uses. They destructively distilled, pulverized, chemically digested, and fermented cobs to produce such basic industrial chemicals as acetone, furfural, charcoal, acetic acid, methanol, oxalic acid, calcium acetate, and formic acid. Their initial results led them to conclude that corncobs might become the source of raw materials for entire industries. Prospects seemed so promising that a *Des Moines Register* reporter visited the campus in 1923, learned of the chemical engineers' research, and gushed that "silk hose, fabric dyes, camera film, 'amber,' stemmed pipes, hot water bottles, chicken feed, and gas mantles are but some of the various objects which can be made from corn cobs."[13]

Despite the appearance of hyperbole, the *Register*'s report was not nearly broad enough. The chemical engineers also used corncob derivatives in the early 1920s for the small-scale production of punk, incense, white lead, vinegar, activated filters for gas works, and water softeners. Demonstrating that chemical engineering could produce a wide range of substances from corncobs created interest in the work, but did not further it. The agricultural waste utilization program rested not on possibilities but on commercial application; the chemical engineers needed to show that isolating a particular substance or manufacturing a specific article would be cost effective and enable an industry to spring up in the state. This stipulation required them to narrow their vision and concentrate on industrial materials most likely to prove commercially successful. It also demanded that they establish in their laboratories small production plants. Only by considering a process from corncobs to final product could they anticipate the costs of large-scale manufacture and compare those expenses with alternative industrial practices.[14]

Iowa State College chemical engineers labored within this utilitarian framework and during this early period devoted most attention to six corncob-based substances. Of these, furfural was the object of greatest experimentation. Prior to the work at Iowa State, furfural remained little more than a laboratory curiosity. Selling for about thirty dollars a pound, this light yellow, oily material's cost made prolonged investigations almost prohibitive. The Ames engineers found, however, that the destructive distillation of corncobs liberated substantial quantities of the substance. They not only refined furfural production techniques and designed blueprints for commercial facilities, but also set about to determine furfural's potential industrial uses. They explored it as a motor fuel, embalming fluid, reducing agent, illuminant, anti-knock additive, source for activated carbon, and base for dyes. Though extensive, these inquiries did not result in the creation of new furfural-based industries. Although the engineers had demonstrated that furfural had many uses, they concluded that in no instance was it sufficiently superior to supplant existing industrial chemicals. More important, the ISC engineers learned as they studied furfural that the Quaker Oats Company had developed a process of producing the substance from oat hulls and was planning to open a production facility at its Cedar Rapids oatmeal plant. While corncob and oat hull furfural production costs were similar, the Quaker Oats factory could produce the substance more economically because it did not have to gather its raw material separately. Corncob furfural would be marketable only if Iowa State's chemical engineers developed new, cheaper production techniques or if a dramatic new use for the chemical caused demand to exceed Quaker Oats' ability to produce it.[15]

The chemical engineers' examination of other corncob derivatives was almost as thorough. After isolating a corncob pentosan, they tried to produce a new plastic from the substance. Detailed tests proved, however, that corncob plastic could not compete physically or economically with available materials. A similar frustration characterized their oxalic acid work. To be sure, oxalic acid had a number of commercial applications—artificial silk, celluloid, dyes, and explosives. But while the Ames engineers demonstrated that they could recover the acid from corncobs, they were unable to manufacture it cheaply enough to interest commercial producers. Xylose, a corncob pentose, also was easy to isolate but held few commercial possibilities. The chemical engineers tried it as an adhesive and as a table-sugar substitute for diabetics, but no investor materialized. The engineers next turned to corncobs as a source of synthetic chicle. Using the heaviest, most resilient portions of the destructively distilled cobs, they took the material to William Wrigley, Sr., to explore its possibilities as a chewing gum base. While Wrigley welcomed the Ames entourage, he complained about the substance's burnt taste. Although Sweeney persisted in chewing it "off and on for about a week," he found that the taste remained strong. Other more conventional attempts to remove the pungent flavor also failed.[16]

The engineers even produced corncob charcoal, targeting it as an inexpensive replacement for wood charcoal in the wood-poor Midwest. The material was cheap and easy to produce, and they demonstrated its versatility by using it to caseharden steel and as a decolorizing agent in sugar refining. They also lobbied Midwestern gunpowder manufacturers to employ it as a substitute for wood charcoal. While not claiming to be "powder experts," the chemical engineers nonetheless mixed their own corncob powder and "succeeded in loading a shell and shooting a rabbit." They then contacted Western Cartridge Company in East Alton, Illinois, presented their case, and waited for the company's reaction. Western Cartridge dismissed corncob char as impractical, but found Iowa State's efforts intriguing. Sweeney seized upon this expression of interest as an opportunity to involve the company in new research on the use of cornstalks. He reminded the company that cornstalks contained a high cellulose percentage that they could convert into smokeless powder—nitrocellulose—rather simply; cornstalks could constitute the raw material upon which to base a segment of the munitions industry. Sweeney's argument apparently impressed Western Cartridge because it provided Iowa State with a grant to fund cornstalk research.[17]

Western Cartridge's grant marked the beginning of Iowa State's sustained cornstalk research. Some preliminary investigations had occurred a few years earlier, and by 1925, the Ames group had established three main processes for converting stalks into fibers: steam under pressure, mechanical manipulation and beating, and chemical digestion. The engineers also had begun to discuss possibilities that cornstalks could become raw material for paper manufacture, cellulose-based explosives, and a synthetic lumber industry. But cornstalk research had progressed slowly before 1925 because of expense; it required special equipment on a large scale. Western Cartridge's grant helped the chemical engineers surmount that hurdle.[18]

Western Cartridge was first to award the chemical engineers outside support, but the engineers were sensitive to the need for publicity much earlier. Almost from their start, they understood that success of their agricultural waste utilization crusade depended as much on promotional activities as it did on engineering efforts. Without creating interest in, excitement about, and markets for their waste-based productions, they could not convince industries to switch raw materials, adopt novel techniques, and reorganize production facilities. Iowa State's engineers publicized their work and stressed its economic benefits to industry whenever possible. Their initial public relations venture took place in 1922 at the Iowa State Fair. Chemical engineering students set up a display and distributed incense, plastic plaques, and other corncob products. They intensified their campaign there in subsequent years. Hoping "to introduce something exotic," the students one year used corncob ingredients to dye "a number of white rats all sorts and kinds of colors." In 1925, cornstalk products joined corncobs. Students constructed their fair booth from cornstalk lumber and decorated the interior with cornstalk tiles. State fairs advertised the chemical engineers' work in Iowa and gave them some national publicity, but the real break came in 1927. Iowa State's engineers were invited to the Eleventh Annual Exhibition of Chemical Industries in New York. Exhibiting hundreds of corncob and cornstalk products, the chemical engineers suddenly burst into national prominence. "New York newspapers featured [our exhibit]," Sweeney later wrote, "and Sunday supplements went all over the United States."[19]

National recognition for Iowa State's chemical engineers paid off in several ways. On the most basic level, they assumed new power within the college's engineering wing. Sweeney's salary was higher than any other department head in the division and more graduate students pursued advanced degrees in his department than in any other engineering discipline. The chemical engineers became a significant voice—perhaps the most significant—in the college's engineering experiment station. Those among them who affiliated with the station outnumbered any other single engineering group by at least two to one. Also, in 1927 the station began to devote an impressively large portion of its annual budget to agricultural waste utilization research and published at least eight separate bulletins on the subject. The Iowa State Engineering Extension Service also fell quickly in line. It issued several popularly written bulletins aimed at farmers and businessmen that featured prominently the chemical engineers' accomplishments. The chemical engineers' new clout extended throughout the college. When the State Board of Education presented its biennial requests in 1926 and 1928, it highlighted the chemical engineers' agricultural waste research and argued that the department's activities justified additional college expenditures. In 1927, the Board had Sweeney address a banquet for the state legislature and discuss his vision of the "greater industrial development" of the state through "intensive utilization of agricultural wastes." The college administration soon were touting the chemical engineers' labors as those "of the very greatest importance to Iowa and the nation."[20]

In the same year that they emerged as a major force in experiment station affairs, the chemical engineers asked for and received a new building expressly for farm-waste investigations. Their new prestige enabled them to equip the facility from outside support. The National Bureau of Standards provided a large grant to outfit the new structure. The Bureau found Sweeney's program so compelling that in the next year it established a research team in Ames to work in tandem with the college's chemical engineers on cornstalk fiberboard production. The USDA Bureau of Chemistry and Soils was also swept up in the excitement in Ames, beginning its own investigations in 1931. The next year it formed a new farm by-product section, and created an agricultural by-products laboratory a year later. The bureau chose to locate the lab at Iowa State. Featuring larger, more modern facilities, the laboratory's staff worked with the college's chemical engineers on destructive distillation techniques, fermentation processes, and cornstalk paper manufacture.[21]

There were other signs that Iowa State's chemical engineers had won a national reputation. In the late 1920s, legislators from states as distant as Georgia were writing to Sweeney and his colleagues for information on their research and its applicability to their states. Foreign governments contacted the chemical engineers. Representatives of Australia, Argentina, Britain, Hungary, and Germany requested information and engineers to establish plants in their countries. The Soviet Union sent two men to Ames in 1929 to examine the engineers' facilities and to learn their practices. American businessmen and innovators, such as Thomas Edison and Henry Ford, regularly corresponded with Sweeney and his cohorts. Sometimes the college's engineers were asked to address industrial concerns. For example, in 1930 General Electric invited Sweeney to discuss how the company could use agricultural waste products in its operations. Among the substances that he proposed were cornstalk lignin as a substitute for shellac, corncob glue, and the employment of pressed cornstalk boards in transformers. Sweeney also found himself representing Iowa State's chemical engineers on prestigious panels. In one instance he shared the platform with such notables as General Motors Chairman Alfred P. Sloan, FBI Director J. Edgar Hoover, and Prime Minister W. L. McKenzie of Canada.[22]

By the late 1920s, Ames's chemical engineers had acquired an international reputation and federal connections. Their research during this period kept pace with their fame. The engineers managed to produce a good quality paper pulp from the stalks and established principles for large-scale cornstalk paper manufacture. In 1929, several regional newspapers dramatized the work by producing cornstalk specials. Newspapers printed on cornstalk paper called attention to the college but were too costly to become anything but a curiosity. The Ames group refined the pulp-making process but was unable to compete with wood paper. Its investigations into alpha cellulose production came to the same end. Although the chemical engineers could make a high-grade alpha cellulose—the base for rayon—from stalks, the cost of the process made commercial manufacture unrealistic.[23]

Other forms of cornstalk pulp seemed more promising. Chemical digestion of stalks in caustic soda yielded a jelly-like pulp that could be molded. Called maizolith—cornstone—the substance was exceedingly dense and a good electrical insulator. Sweeney and his crew projected it as "knife handles, switch buttons, knobs, gears," and other high-stress objects. They were even more optimistic about the pulp-like material that resulted from attacking cornstalks with pressurized steam. This material could form synthetic lumber of varying densities. Productions ranged from substitutes for cork, which initially interested refrigerator manufacturers, to a hard, dense board suitable for "the construction of automobile bodies, truck bodies, [and] Pullman cars." The pressurized steam process possessed an additional advantage. Its waste water contained a high percentage of organic matter which, when collected, could serve as a source of raw materials for other industries.[24]

Perhaps because it produced two commercial substances from one basic process, the chemical engineers emphasized cornstalk board manufacturing. They focused on cornstalk insulation board, a choice that reflected commercial realities. The extant insulation board industry was a synthetic lumber industry. The two most common insulation boards, Insoboard and Celotex, were composed of straw and sugar cane bagasse, respectively. With an already established market for synthetic insulation lumber, Ames's chemical engineers only needed to produce cornstalk insulation board more cheaply than competitors. In this hope the Iowa State group worked in two complementary directions. The engineers explored manufacturing and drying techniques for cornstalk insulation board, and also desperately sought commercial uses for the liquor. The latter endeavor was far-ranging. The chemical engineers examined cook liquor as a source of natural gas, flotation oils, oxalic acid, lignin, xylose, polyhydroxyl alcohols, and adhesives. They even tried it as a road binder.[25]

The chemical engineers did more than experiment; they also tried to involve industry. Sweeney was particularly aggressive and almost immediately successful. In 1927 a group of Chicago businessmen established Maizewood Products to manufacture cornstalk insulation boards. Sweeney served as informal consultant to the firm and lent his considerable prestige to its effort. His business acumen, unfortunately, did not match his scientific talents, however. Construction on the production facility in Dubuque began in 1927, but because of financial difficulties was not complete until late in 1929. Almost simultaneously with the plant's opening and production of the first boards—they came four feet wide, seven-sixteenths of an inch thick, and eight, nine, or ten feet long—the company went into receivership. It had exhausted its limited funds.[26]

The receivers sought new ownership and prevailed on Sweeney to take complete charge of production while the search was underway. He agreed, and his reputation as well as, by implication, the reputation of Iowa State's chemical engineers, soon paid dividends. Their faith in cornstalk insulation board had attracted several prominent Iowans who in 1930 established a holding company, National Cornstalk Processes, to take over Maizewood. The investors who provided the then impressive sum of one million dollars included Henry A. Wallace, soon to become secretary of agriculture; Frank O. Lowden, former governor of Illinois; and Herbert E. Perkins, president of International Harvester. The new owners moved swiftly; they contacted Sweeney and asked him to reorganize the Dubuque plant. Sweeney revamped the established production techniques and added new ones; he employed the latest processes which Iowa State's chemical engineers had uncovered, some of which they had not yet even patented. His work produced dramatic results. By 1932, Maizewood was turning out roughly seventy thousand feet of cornstalk insulation board per day. Among its clients was the Chicago World's Fair. Maizewood insulating board was used in the fair's administration building and was featured prominently, which publicized both the firm and Iowa State.[27]

Cornstalk insulation board was the first modestly successful agricultural waste product devised at Iowa State. Yet even before Maizewood, the chemical engineers had demonstrated the manifold possibilities of farm wastes and the excitement of their labors had rubbed off on their colleagues. Members of other departments had gravitated to the agricultural waste program. Several had become active, if occasional, participants. For instance, in 1928 the agricultural experiment station's J. B. Davidson and E. V. Collins developed a cornstalk harvester. R. M. Hixon and C. J. Peterson of the college's chemistry department the next year investigated the chemical composition of cornstalk tissue. In 1929 members of the bacteriology department studied the bacteria instrumental to the fermentation of cornstalks. Each effort was independent of the chemical engineers, but agricultural waste utilization research became an activity around which a substantial portion of the faculty rallied. During the 1930s it would usher in an unprecedented period of interdepartmental cooperation and coordination as faculty from different departments and wings worked together to resolve technical issues. Farm waste investigations were no longer simply the province of chemical engineers, but of the college generally.[28]

Acceptance of agricultural waste research as a subject that required interdepartmental action broke down outdated jurisdictional barriers within the college and spurred communication among disciplines. A prime example of the new trend was a college-sponsored exploration of agriculturally derived "alcohol and other chemicals" as motor-fuel additives. In 1932 and 1933 the "Iowa State College Committee On the Use of Alcohol in Motor Fuels" conducted the research as an ad hoc body of eleven faculty members who represented four departments: chemistry, mechanical engineering, agricultural economics, and chemical engineering. The manner in which committee members approached their investigations was decisive and revealing. Unlike the chemical engineers' earlier efforts—in which they considered an entire industrial process from gathering the waste to establishing production facilities and techniques—committee members carefully divided tasks. Each department examined only those portions of the larger question that related to its special expertise. For instance, mechanical engineers performed mileage tests on alcohol-gasoline and acetone-gasoline blends. They also measured power output, influence on starting, and anti-knock values. Chemists' responsibilities were similarly circumscribed. They surveyed the literature for other alcohol-gasoline tests, determined the miscibility of several alcohol-gasoline blends, and prepared various test mixtures. In addition, they compared the costs of producing alcohol from corn with that of producing it from a starch-grain-molasses equivalent. This type of research design was new at the college. Predicated upon the interdependence of faculty members, it defined "colleague" in a different way, as cross-disciplinary and functional. No single department dominated within this structure because each brought a particular skill that seemed necessary for the problem at hand. Thus this interdisciplinary research group was organized not hierarchically but functionally. Its existence signified the beginning of a new relationship between the college's previously distinct wings.[29]

The chemical engineers' college transformation received an assist from Raymond M. Hughes, who assumed the presidency of Iowa State College in September 1927. Hughes set the tone for interdisciplinary research by creating a council on research and forming as council subcommittees the Agricultural and Engineering Experiment Stations, the Industrial Science Research Institute, and the Veterinary Research Institute. Hughes's administrative act made clear that he considered research a common enterprise, unique among college activities. It divided the faculty into two segments, researchers and others, and that new explicit commonality gave researchers justification for interdepartmental and interdivisional cooperation.[30]

The initial response to Hughes's realignment scheme had been far from encouraging. Only a few faculty members engaged in even the most modest cooperative ventures. Professors in agriculture and industrial science studied land utilization patterns and formulated objectives for rural life. Also, within the agricultural experiment station, a social science section and a rural education subsection appeared. Although these efforts were interdepartmental and interdivisional, the rearrangement that Hughes had hoped to foster did not immediately occur collegewide. Interdivisional permutations included only agriculture and industrial science. Veterinary medicine and especially engineering continued to reside outside the new cooperative research nexus. Therefore, making the coordinated research scheme a college-wide reality stood as the chemical engineers' major institutional achievement. To be sure, their significance also extended to other areas. They had made chemical engineering important within the college and throughout the nation, and they showed the way for more broad-based, problem-oriented cooperative research endeavors. Their agricultural waste utilization program of the 1930s was the first instance in which engineering research emerged as an integral part of the college's new interdivisional program. In effect, they bridged the chasm that had distinguished engineering and agricultural wings in a new manner. While in the 1920s the chemical engineers had chosen utilization of agricultural wastes as their primary industrial mission, it remained very much a department-based initiative. Only after 1930 did it lead to interdivisional action between engineering and agriculture faculty. Then, in the later 1930s, interdivisional problem-oriented research teams became de rigueur throughout Iowa State. The college's unique Corn Research Institute and its Swine Breeding Laboratory stand as the best-known examples.[31]

Hughes certainly did his part to speed the diffusion of research teams throughout the college. Shortly after the project to make motor fuel from alcohol began, he addressed the faculty. He lauded interdivisional, problem-oriented research groups and celebrated the chemical engineers' role in creating the school's first such effort. He applauded the motor fuel study as an instance in which "departmental and divisional lines have been ignored" and he maintained that the practice had unleashed a "new institutional power . . . , the power of Iowa State College as a whole." He called on the college staff to replicate the process and conceive of Iowa State as a single "working unit." Although he acknowledged that "divisions and departments are necessary administrative units," Hughes concluded that they "are artificial." His explanation of the need for cooperation was simple: "Problems know no departmental lines." Their resolution usually required the union of "two or more departments, regardless of divisions," which would "work together from different angles and in frequent consultations."[32]

Ironically, the chemical engineers' heyday ended just when they had achieved their greatest institutional significance. Two radically different problems occasioned their downfall. Their narrowly focused research program—corncobs in the early 1920s, cornstalks thereafter—had resulted in precious few commercial applications. Only the Maizewood process seemed commercially viable, and that viability became questionable in the mid-1930s when the Depression struck the housing industry particularly hard, and federal financial support also became rare. The Bureau of Standards removed its research team in 1933, and the USDA relocated its agricultural by-products laboratory to Peoria, Illinois, a few years later. Even before the federal government withdrew support, there were indications that the chemical engineers recognized their plight. This realization did not cause them to abandon the idea that agricultural wastes could serve as industrial raw materials. Instead, they continued to follow, metaphorically if not literally, Sweeney's maxim "that there is more cash value in the elaborated stalk and cob than there is in the corn." Perhaps as an acknowledgement that the future of their waste utilization campaign hinged upon the immediate development of a commercially feasible product, the Ames group tried such varied agricultural wastes in the 1930s as barley, wheat, flax, rice, and oat straw; cotton, banana, popcorn, broomcorn, and tobacco stalks; rice, oat, and cottonseed hulls; swamp root; peanut and pecan shells; pineapple and Jerusalem artichoke tops; milkweed plants, tobacco stems, and sorghum cane bagasse; beet-sugar waste; and reeds, peat, and even chicken feathers. Yet while the chemical engineers considered a broad range of wastes throughout the decade and devoted special attention to cornstalk insulation board, furfural, and soybean oil, their new agricultural waste investigations did not bear fruit and their influence was on the wane. They had patented nineteen devices or processes from 1927 to 1935, but only one from 1938 to 1950.[33]

Since they had not developed new commercially applicable processes and techniques from agricultural wastes, the chemical engineers lost their influence with industry. But the reason for their declining power within the college was far subtler. The chemical engineers as a cohesive department, organized around the agricultural waste utilization campaign, fell victim to their own achievement, the interdepartmental, interdivisional research groups. Cooperative, problem-oriented research groups had transformed departmental coherence from a benison to an anachronism; departments focusing on only one arena, such as chemical manufacture from farm waste, had little utility within the new scheme. Problem-oriented research groups depended on specialists from different divisions and departments who would come together temporarily and devote their expertise to the particular question. Departmental coherence or homogeneity, then, was atavistic. Members of such departments could contribute to only a small fraction of college research activities. For a department to blossom or remain preeminent within this context, its faculty had to possess a wide range of specialties and thus be able to participate in many research groups. Only through the participation of a department's individual members on numerous research teams could the department stand tall within the "new" college.

The chemical engineers found themselves between a rock and a hard place. To regain industrial support required development of new commercial agricultural waste utilization processes and techniques, but to recapture primacy in the college they would have to broaden their interests significantly to cooperate with many research groups. Before World War II the chemical engineers attempted to do both and accomplished neither. In 1937, and for the first time during Sweeney's tenure at Iowa State, only a minority of students selected farm waste thesis topics. This trend persisted through 1941, but the apparent shift away from almost exclusive emphasis on agricultural waste research did not give the chemical engineers immediate entry into the reconstituted college community. Department members virtually disappeared from engineering experiment station and extension service publications. Between 1938 and 1949, the chemical engineers contributed only a single bulletin.[34]

World War II interrupted the chemical engineers' search for a renewed mission. They did not reestablish industrial ties or become integrated within the college. The pace of agricultural waste work quickened somewhat during this period, but it differed strikingly from earlier research; it shunned long-range prospects and focused on immediate benefits. Its goal was not to boost farm incomes but, in a period marked by shortages and rationing, to turn readily retrievable wastes into useful products. Rather than consider conventional agricultural commodities, the chemical engineers attempted to derive industrial substances from temporary, war-related crop wastes. Hemp and milkweed refuse received the most attention. Japanese occupation of the Philippines had cut off Americas supply of hemp rope, while demand for milkweed floss sharply increased because of its use in airplane acoustical tiles. Hemp-hurd research came first. The chemical engineers proved they could produce insulation boards of passable quality from hurds. They even prevailed on Maizewood to manufacture some to see how the process functioned on an industrial scale. Only later in the war did they take up milkweed seed. After a number of tests, they had difficulty finding a use for the material. The war concluded before they resolved the problem. With the war at an end, they dropped their attempts to develop an application for the milkweed seeds and turned away from hemp-hurd investigations.[35]

After the war, the chemical engineers' research efforts became much more diverse. Only Sweeney and Arnold continued agricultural waste investigations, and they limited their inquiries sharply to cornstalk insulation board, furfural, and soybean oil. Each of the department's other members went in a different direction. Even Webber abandoned agricultural wastes and specialized in unit operations. Sweeney resigned as department chairman in 1947; his department had outgrown his leadership. It had chosen to become a department of many specialties and part of the integrated college.[36]

Any assessment of Iowa State's chemical engineers during the interwar period must conclude that their agricultural waste utilization campaign paid farmers and industrialists few immediate benefits. Farm wastes did not serve in the 1920s and 1930s as the source of raw materials of entire industries; kernels remained more valuable than cobs and stalks. Despite this failure, the chemical engineers had accomplished much institutionally. Their research helped put Iowa State College on the map. It gave the school its initial international reputation and attracted investment from the federal government and private enterprise. Their research program also served as a base around which to organize the college faculty. In the 1920s, it bridged the gap between Iowa State's distant engineering and agricultural wings, and its contributions were even greater in the next decade. The chemical engineers paved the way for the creation of interdisciplinary and interdivisional research groups. Institutional reformation was no small matter. In this sense the chemical engineers did indeed aid farmers, though not immediately and not in the manner proposed. While their agricultural waste investigations did not boost farm incomes, the chemical engineers' cooperative research served as a model by which the college faculty could tackle large multidimensional problems. After the chemical engineers' path-breaking efforts, the college repeatedly formed interdisciplinary, interdivisional research teams. Today they remain the heart of the modern university.

Notes

This essay appeared in substantially the same form in the *Annals of Iowa*. We thank them for their permission to use it here.

1. For the situation at Iowa State, see Erik Lokensgard, "Formative Influences of Engineering Extension on Industrial Education at Iowa State College" (unpublished dissertation, Iowa State University, 1986). Also see R. A. Pearson, "Address to Grant Club, February 20, 1913," Raymond A. Pearson Papers, Special Collections, Iowa State University Library; Earle D. Ross, *A History of the Iowa State College of Agriculture and Mechanic Arts* (Ames, 1942); Barton Morgan, *A History of the Extension Service of Iowa State College* (Ames, 1934), 24–33; Christie Dailey, "Implementation of the Land-Grant Philosophy during the Early Years at Iowa Agricultural College, 1859–1890" (M.A. thesis, Iowa State University, 1982), 21–29; and Iowa State Board of Education, Biennial Report, 1912, 8–10, 335–339. For the situation at some other land-grant colleges, see Morris Bishop, *A History of Cornell* (Ithaca, 1962); Monte A. Calvert, *The Mechanical Engineer in America, 1830–1910* (Baltimore, 1967), 87–105; Merle Curti and Vernon Carstensen, *The University of Wisconsin, 1848–1925: A History*, 2 vols. (Madison, 1949), 1: 459–475, 501–533, 2: 374–410, 444–467, and

Richard P. McCormick, *Rutgers: A Bicentennial History* (New Brunswick, 1966), 89–94, 118–122, 153–155, 174–185.

2. Minute Books of the Iowa State College Faculty, 6 May 1913.

3. Charles A. Mann, "Chemical Engineering Course at Ames," *Iowa Engineer* 17 (1916), 86–87; and Iowa State Board of Education, Biennial Report, 1914, 247; and 1918, 236. Undergraduate theses included E. R. Barnum, "The Recovery of Paper Mill Waste in the Form of White Portland Cement," 1915; R. S. McMullin, "A Study of the Condensation Products Formed in the Manufacture of Carburetted Water Gas," 1916; Harold P. Roberts, "The Elasticity and Tensile Strength of Paint Film," 1916; Hobart J. McKay, "Manufacturing Alcohol from Corncobs," 1916; Charles E. Ricker, "Production of Sodium Cyanide Utilizing Nitrogen of the Air," 1917; Homer C. Clarke, "The Extraction of Iron and Aluminum Sulphates from the Lakoata Coal Washery Dump," 1918; Leon C. Heckert, "The Formation of Toluene from Benzene-Xylene Mixture by the Use of the Friedel Crafts Reaction," 1918; A. L. McMillan, "Absorption of Carbon Monoxide from the Air by Charcoal under Atmospheric Conditions," 1918; and A. W. Pickford, "The Corrosion of Iron," 1918. Graduate theses included Werner J. Suer, "A Study of the Carbohydrates of the Corn Cob," 1917; and George H. Montillon, "The Production of Acetone from Corncobs," 1917.

4. For celluloid, see Robert Friedel, *Pioneer Plastic: The Making and Selling of Celluloid* (Madison, WI, 1983). For the importance generally in early-twentieth-century American thought, see Jeffrey L. Meikle's masterful *American Plastic: A Cultural History* (New Brunswick, NJ, 1995). Also see Stephen Fenichell, *Plastic: The Making of a Synthetic Century* (New York, 1996); and Penny Sparke, ed., *The Plastics Age: From Bakelite to Beanbags and Beyond* (Woodstock, NY, 1993).

5. The story of German chemistry is told concisely in John Joseph Beer, *The Emergence of the German Dye Industry* (Urbana, IL, 1959).

6. The situation is recounted in the definitive court case; see U.S. Supreme Court, *United States v. Chemical Foundation, Inc.*, 272 U.S. 1 (1926).

7. Iowa State Board of Education, Biennial Report, 1916, 227, 231-234, 245–249, 254-255, 292-294; Ross, *History*, 274-287, 326-330.

8. "Prominent Chemical Expert Will Head Department Here," *Iowa State Student*, 29 September 1920, 5; "He Invented an Industry," *The Milepost* 2 (16 October 1930), 27.

9. O. R. Sweeney and H. A. Webber, "Experimental Studies on the Destructive Distillation of Corncobs," Iowa Engineering Experiment Station Bulletin #107 (Ames, 1931), 6; Sweeney to T. R. Agg, 28 January 1936, Orland R. Sweeney Papers, Special Collections, Iowa State University Library; Sweeney to S. W. Smith, 13 April 1934, Sweeney Papers; Harry S. Morgan, "Molders of Men— O. R. Sweeney," *Iowa Engineer* 38 (1938), 170.

10. *Iowa State Student*, 29 September 1920; O. R. Sweeney, "Chemical Engineering," *Iowa Engineer* 21 (May 1921), 5; O. R. Sweeney, "Utilization of Agricultural Products," speech draft, 3 December 1937, Sweeney Papers; "Cornstalk

Cellulose Industry Assumes Gigantic Proportions," *Current Topics* (1928), 3; O. R. Sweeney, "Possibilities in Manufacturing By-Products of Corn," *Iowa Yearbook of Agriculture,* 1926, 21–22.

11. A typescript list of chemical engineering theses done prior to 1933 is available at the Iowa State University Library. See Lionel K. Arnold, comp., "An Annotated Check List of Chemical Engineering Theses from Iowa State College" (39 pp.) and "Supplementary Check List of Chemical Engineering Theses" (8 pp.).

12. For rosters of the chemical engineering staff, see the following pages in the biennial reports of the Iowa State Board of Education: 1922: 210; 1924: 185; 1926: 253; 1928: 318; 1930: 324, 341; and 1932: 279. Aside from Arnold and Webber, no chemical engineer remained a member of Sweeney's department for more than four years.

13. Chemical engineering theses from these initial research projects include J. A. Hovsepian, "Studies on Pentosan Plastics," 1924; James M. Williamson, "Production of Plastics from Oak Hulls," 1923; Ted Bergman, "Utilization of Beet Sugar Pulp," 1921; Verne Hass and Tom Gilbert, "The Utilization of Sugar Beet Pulp," 1922; E. J. Mleynek, "The Recovery of Pectin from Fruit Wastes," 1923; and H. L. Shepard and H. A. Howell, "Auto-Catalysis in the Hydrogenation of Cotton Seed Oil," 1921. For work on corncob utilization, see O. R. Sweeney, "The Commercial Utilization of Corncobs," Iowa Engineering Experiment Station Bulletin #73 (Ames, 1924) and the following chemical engineering theses: G. W. Burke, "Some Analytical Data On Corn Cobs and Their Parts," 1923; L. K. Arnold, K. L. Wagner and H. G. Goldschmidt, "A Study of the Destructive Distillation of Corncobs," 1921; G. P. Patterson and H. R Bigler, "Destructive Distillation of Corn Cobs," 1922; L. M. Christensen, "A Study of the Utilization of Corn Cobs by Fermentation," 1923; George M. Russell and Harvey C. Morris, "The Production of Vinegar," 1922; L. J. Botleman and R. R. Wagner, "Preparation of Furfural From Corn Cobs On a Commercial Scale," 1921; and Richard L. Hanson and Jack W. Hussey, "A Study of the Production of Calcium Acetate from Corncobs," 1923. See also Ruth Cromer, "How to Make Money from Iowa Corn Cobs," *Des Moines Register,* 7 October 1923; "Chemists Create Useful Materials from Corncobs" and "Lowly Corncob Furnishes Many Unsuspected Daily Products," *Iowa State Student,* 15 April 1921, 1 and 24 January 1921, 3.

14. Relevant chemical engineering theses include R. P. Moscrip, "Studies in Corn Cobs," 1922; George T. Williams, "The Commercial Utilization of Corncobs," 1923; Kenneth M. Vaughn, "Utilization of Corncob Alkali Fusions in Water Softening," 1926; and Robert E. Fothergill and Ronald L. McVey, "A Study of the Methods for Reducing Corncobs to Various Degrees of Fineness and the Utilization of the Ground Material," 1924.

15. Chemical engineering theses useful on these points include J. P. Bishop and D. L. Gilbert, "The Preparation of Furfural From Corncobs," 1921; L. N. Haugen and E. J. Kowalke, "Further Investigations on the Commercial Production of Furfural From Corn Cobs," 1923; Galen Hunt, Philip C. Jones, and Thalmer J. Thompson, "The Preparation of Activated Carbon From Furfural Residues,"

1924; H. M. Wolcott, "Possible Use of Furfural as a Motor Fuel," 1923; and Paul Bruins, "Commercial Utilization of Agricultural Wastes," 1927. See also Sweeney's retrospective comments on furfural and Quaker Oats: Sweeney to Elizabeth L. MacHatton, 11 November 1943; Carl S. Miner to Sweeney, 26 February 1946; and Sweeney to Miner, 1 March 1946, all in Sweeney Papers.

16. Relevant chemical engineering theses on conversion to plastic include Herbert L. Bowers, "A Study of Synthetic Resins from Pentosan Materials," 1924; Albert J. Duden, "Properties of Corn Cob Plastics," 1925; Joseph A. Hovsepian, "Studies on Pentosan Plastics," 1924; and J. E. McFarland, "Plastic Condensation Products From Pentosan-Containing Materials," 1924. Theses on oxalic acid included Ralph M. Cash and Gaylen Sayler, "Preparation of Oxalic Acid from Corn Cobs," 1922; Raleigh L. Farlow, "Preparation of Oxalic Acid from Corncobs," 1923; Ralph A. Trexel, "Further Studies on the Production of Oxalic Acid from Corn Cobs," 1923; and Henry A. Webber, "Studies on the Production of Oxalic Acid from Corn Cobs and Stalks," 1925. Theses on xylose included D. R. Kiser, "The Manufacture of Crude Xylose from Corncobs," 1925; W. E. Rouser, "Manufacture of Xylose from Corncobs," 1925; and Julian Toulosue, "Xylose from Corn Cobs," 1926. Sweeney discussed the chicle work retrospectively in Sweeney to L. W. Marble, chief chemist, Frank H. Fleet Corporation, 19 July 1939, and Sweeney to Robert L. Wilson, director of research, William Wrigley Company, 6 August 1942, both in Sweeney Papers.

17. Sweeney, "Possibilities in Manufacturing" (see n. 9), 25; Sweeney to H. F. Smith, Smith Gas Engineering Company, 14 May 1926, Sweeney to Western Cartridge Company, 4 September 1925, both in Sweeney Papers. Sweeney summarized much of the early corncob work at Iowa State in "The Commercial Utilization of Corncobs," Iowa Engineering Experiment Station Bulletin #73 (Ames, 1924). See also Lionel K. Arnold, *History of the Department of Chemical Engineering at Iowa State University* (Ames, 1970), 25.

18. Chemical engineering theses that considered furfural during this period included Paul R. Bruins, "Application of Furfural and Its Derivatives to Manufacturing Plastics," 1930; Richard W. Bruins, "The Physical Properties of a Furfural Plastic," 1930; Jacob D. Green, "Pyrolysis of Furfural," 1931; Lyle K. Huhn, "Studies on the Products from Furfural Cracking," 1931; Robert V. Janda, "The Cracking of Furfural," 1931; and Lawrence G. Mason, "A Study of the Action of the Products Obtained from the Maize Plant in Rubber Compounds as Fillers, Softeners, Accelerators and Agars," 1928. Early chemical engineering cornstalk utilization theses included Richard Ericson, "Experiment with Furfural Yields from Corncobs and Cornstalks," 1925, and Oscar Tow, "The Production of Paper from Iowa Cornstalks," 1925. Also see Lionel K. Arnold, "Cornstalks as an Industrial Raw Material," *Agricultural Engineering* 9 (1928), 379–380, and "Making Insulating Boards from Cornstalks," *Cellulose* 1 (1930), 272–275.

19. Sweeney recounted public relations efforts in Sweeney to H. E. Barnard, director of research, National Farm Chemurgic Council, 7 May 1940, Sweeney Papers.

20. Iowa State Board of Education, Biennial Report, 1926, 253, and Biennial Report, 1928, 285. The engineering experiment station published a list of staff members with each bulletin. The chemical engineers never had less than four people on staff, while no other engineering division had more than two. For the station bulletins, see O. R. Sweeney and L. K. Arnold, "Cornstalks as an Industrial Raw Material," #98, 1930; "The Production of Paper from Cornstalks," #100, 1930; and "Studies on the Manufacture of Insulating Board," #136, 1937; O. R. Sweeney, C. E. Hartford R. W. Richardson, and E. R. Whittemore, "Experimental Studies on the Production of Insulating Board from Cornstalks," #102, 1931; O. R. Sweeney and H. A. Webber, "Experimental Studies on the Destructive Distillation of Corncobs," #107, 1931; H. A. Webber, "The Production of Oxalic Add from Cellulosic Agricultural Materials," # 118, 1934; L. K. Arnold, H. J. Plagge, and D. E. Anderson, "Cornstalk Acoustical Board," #137, 1937. Extension service bulletins included L. K. Arnold, "Utilization of Agricultural Wastes," #99, 1928, and "Utilization of Agricultural Wastes and Surpluses," #113, 1933; and O. R. Sweeney, J. H. Arnold, and L. K. Arnold, "Processing the Soybean," # 103, 1929.

21. "Cyrenus Cole on the Utilization of Agricultural Wastes," speech delivered to the U.S. House of Representatives, 25 January 1927, Sweeney Papers; Sweeney to S. A. Knapp, 13 May 1936, Sweeney Papers; Virgil W. Anderson, "Ahead of the Field," *Iowa Engineer* 37 (1937), 127; Warren E. Emley, "Processes for Industrial Use of Corn Stalks Are Developed," *United States Daily*, 21 January 1931, 7; Iowa State Board of Education, Biennial Report, 1926, 21, and Biennial Report, 1928, 56–61; O. R. Sweeney, "Greater Industrial Development of Iowa by More Intensive Utilization of Agricultural Wastes," banquet address delivered to officials and legislators, 18 February 1927, Sweeney Papers; R. M. Hughes, Address... September 17, 1930, 29; C. E. Hartford, "Manufacture and Properties of a Cellulose Product (Maizolith) From Cornstalks and Corncobs," National Bureau of Standards Miscellaneous Publication #108, 1930; O. R. Sweeney and W. E. Emley, "Manufacture of Insulating Board From Cornstalks," National Bureau of Standards Miscellaneous Publication #112, 1930; George K. Burgess, director, Bureau of Standards, "Annual Report, 1930," National Bureau of Standards Miscellaneous Publication #115, 1930, 44–45; Lionel K. Arnold, "Chemical Engineering Building at Iowa State College," *Journal of Chemical Engineering* 6 (1929), 292–298; USDA Bureau of Chemistry and Soils, Report of the Chief, 1928, 30; 1930, 2; 1932, 14; 1934, 8. For the bureau's research see H. G. Knight, "Industrial Utilization of Farm Products," *Condensed Proceedings of the Southern Chemurgic Conference* 69 (1936), 8–14; H. T Herrick, "The Research Program of the Bureau of Chemistry and Soils on Industrial Utilization of Farm Products," *Proceedings of the American Soybean Association*, 1937, 3–9; and T. R. McElhinney, B. M. Becker, and P. B. Jacobs, "Destructive Distillation of Corncobs: Effect of Temperature on Yields of Products," *Industrial and Engineering Chemistry* 30 (1938), 697–701.

22. "Russian Government Attracted by Cornstalk Process," *Iowa Engineer* 29 (1919), 26; Sweeney to Mrs. R. C. Good, 2 October 1929; Eugene Matrosow to Sweeney, 20 May 1930; W. W. Larsen to Sweeney, 20 May 1930; Sweeney to E.

G. Liebold, general secretary to Henry Ford, 15 February 1933; Sweeney to Professor W. H. Stevensont 15 July 1931; Sweeney to T. R. Agg, 25 November 1930; all in Sweeney Papers. "Dr. Sweeney Speaks to Insurance Presidents," *Iowa Engineer* 38 (December 1937), 66. Sweeney's letter to Liebold discusses his previous conversations with Ford and Edison.

23. Theses on paper pulp from cornstalks include Lionel K. Arnold, "The Elaboration of Corn Stalks into Paper Pulp," 1926; Lionel K. Arnold, "Cornstalks as a Raw Material for Paper Production," 1930; Stanley J. Hultman, "Cornstalk Pulp by the Sulphite Process," 1930; Ralph H. Riemenschneider and Stuart M. Garstren, "Studies in the Manufacture of Paper from Cornstalks," 1926; William Roberts, "A Study of the DaVains Process for Making Paper from Cornstalks," 1927; Chitao P. H. Tan, "Corn-Stalks as a Raw Material for Paper Manufacture," 1926; William Perry Wood, Jr., "Some Chemical and Physical Properties of Various Cornstalk Papers," 1928. For some cornstalk specials, see the *Webster City Daily Freeman-Journal*, 28 March 1929; *Spencer News-Herald*, 28 February 1929; and the *Huron* (South Dakota) *Evening Huronite*, 10 January 1929. Also see Sweeney to Frank E. Deem, A. J. Brandt, Inc., 4 April 1931; Sweeney to E. Jerome Dyer, London, 30 March 1931, both in Sweeney Papers; Lionel K. Arnold, "Cornstalks as a Papermaking Material," *Paper Trade* 53 (1927), 3–5, and "Paper from Cornstalks," *Cellulose* 1 (1930), 224–227. On cellulose see H. A. Webber, "Cellulose from Cornstalks," *Industrial and Engineering Chemistry* 21 (1929), 269–271. Theses included Clark Bright, "A Study of the Methods of Preparing High Alpha Cellulose from Cornstalks," 1928; Frank P. Fowler, "Nitric Acid Hydrolysis of Corn Stalks," 1928; Charles A. Funk and Carrell O. Turner, "Studies on the Isolation of Cellulose from Corn Stalks by Chlorination," 1929; and Darwin H. Huff, "Studies on the Isolation of Corn Stalk Cellulose with Dilute Nitric Acid," 1929.

24. Arnold, "Utilization of Agricultural Wastes" (see n. 20), 8. Maizolith theses by chemical engineers included Jack W. Eichinger, "The Production of Cellulith from Cornstalks and Corncobs," 1926; Charles Hartford, "The Manufacture of Cellulith from Cornstalks and Corncobs," 1928; Ronald J. Berkhimer, "The Waterproofing of Maizolith," 1929; Jacob D. Matlack and John M. Sharf, "Drying and Molding of Maizolith," 1931. Synthetic lumber theses included John S. Reece, "Synthetic Lumber from Cornstalks," 1928; Paul A. Leightle, "Miscellaneous Studies on Water Proofing of Synthetic Lumber from Cornstalks," 1929; and J. L. McMurphy, "The Sizing and Waterproofing of Press Boards Made from Cornstalks," 1930. Also see O. R. Sweeney and Robley Winfrey, "The Production of Synthetic Lumber from Cornstalks," *Mechanical Engineering* 52 (1930), 848–851.

25. Arnold, "Utilization of Agricultural Wastes," 7; Sweeney, "Possibilities in Manufacturing By-Products of Corn" (see n. 10), 23–24. Theses on cornstalk lumber included Roy A. Loomer, "The Production of Synthetic Lumber from Corn Stalks and Plaster of Paris," 1927; Murray B. Peterson, "Apparatus for the Control of the Process Used in the Manufacture of Synthetic Lumber," 1929; Hubert C. Richardson, "The Production of a Thermal Insulating Material of

Low Conductivity from Corn Stalks," 1929; Robert W. Richardson, "Development of Synthetic Lumber from Cornstalks," 1930; G. M. Seidel, "The Production of Synthetic Lumber from Cornstalks," 1927; and Herbert L. Stiles, "The Thermal Conductivity of Corn Stalk Products," 1929. Cook liquor theses included Willett J. McCortney, "Composition of Cornstalk Cook Liquors," 1927; Norman F. Kruse and James A. Shea, "The Recovery of Lignin from Cooked Cornstalk Liquor," 1929; Theodore R. Naffziger, "Plans and Specifications of a Plant for the Manufacture of Adhesive from the Wastes of the Corn Plant," 1928; and James W. Beach, "Cook Liquor as a Core Binder," 1931.

26. Sweeney to Herman Manufacturing Company, Lancaster, Ohio, 18 November 1929, Sweeney Papers.

27. J. S. Russell, "Magnates Back Cornstalk Mill," *Des Moines Register,* 12 January 1930, 19; Sweeney to Blackhills Mining and Industrial Association, Rapid City, South Dakota, 10 September 1931, Sweeney Papers; O. R. Sweeney's letter to the editor, *Saturday Evening Post,* 5 April 1940.

28. James E. Franken, "The Middlewest's Own Industry: Cornstalk Wallboard," *Iowa Engineer* 29 (1928), 3–4; J. B. Davidson and E. V. Collins, "Harvesting Cornstalks for Industrial Uses," Iowa Agricultural Experiment Station Bulletin #274 (Ames, 1930); C. J. Peterson and R. M. Hixon, "Chemical Examination of the Tissue of the Cornstalk," *Industrial and Engineering Chemistry, Analytical Edition* 1 (1929), 65. Two bacteriology master's theses emerged during this period: Roger Patrick, "Bacteria Fermenting Xylan," 1929, and R. H. Carter, "Thermophilic Fermentation of Carbohydrates," 1929.

29. For the division of responsibilities and plan for investigation see the foreword to "Use of Alcohol as Motor Fuel-Iowa State College," typescript, Iowa State University Library. This volume contains seven papers that constitute the results of the investigation.

30. R. M. Hughes, Address ... September 23, 1927, 6–8; and September 22, 1928, 4, 6, 10. Also see Ross, *A History,* 357.

31. R. M. Hughes, Address... September 16, 1931, 17–18; C. F Curtiss, "Report on Agricultural Research," *Iowa Year Book of Agriculture,* 1932, 145–147.

32. R. E. Buchanan, "Report on Agricultural Research," *Iowa Year Book of Agriculture,* 1934, 353–355, 358–359; R. M. Hughes, Address... September 18,1935, 6, 24–25; Charles E. Friley, President's Address September 14, 1938, 2–3; Iowa State Board of Education, Biennial Report, 1934, xxxi–xxxii, and 1936, li–liv.

33. R. M. Hughes, Address September 20, 1933, 3–4, and Address September 19, 1934, 10; W. L. Bierring to Sweeney, 23 May 1938, and Sweeney to Fred E. Butcher, 8 July 1943, both in Sweeney Papers. Sweeney's comment was reported in Wheeler McMillen, "Uses Now Found for Agricultural Waste," *New York Times,* 2 October 1927, 14. For brief reports of the chemical engineers' broad-gauged research in the early 1930s, see Sweeney to L. J. Dickinson, 10 December 1931; Sweeney to E. Jerome Dyer, 9 July 1939, 26 July 1937, 30 March 1931; and Sweeney to Waldemar Kaempfert, 16 July 1934, all in Sweeney Papers. See also USDA, Bureau of Chemistry and Soils, Report of the Chief, 1932, 14; and Arnold, "Utilization of Agricultural Wastes and Sur-

pluses," 9. Representative theses from the 1930s included P. N. Burkhart and Dan M. Harrison, "Studies in the Extraction of Soybean Oil with N-Butyl Alcohol," 1937; H. E Conway and J. J. Wayler, "Furfural as a Paint and Varnish Remover," 1937; R. B. Menzel and P. E. Seeling, "The Destructive Distillation of Alkali Lignin," 1937; Carl A. Holmberg, "The Effect of Composition on the Drying Rate of Corn Board," 1939; James Dustin and Harold Heap, "Emulsions of Furfural and Their Uses," 1938; and Arnold L. Ayres, "A Soybean Furfural Urea Plastic," 1938. The only patent received between 1936 and 1950 by a chemical engineer was O. R. Sweeney, Matboard Handing Machine, 1937, US Patent Office #2,084,980.

34. In 1937, only fifteen of thirty-four students selected thesis topics concerning agricultural wastes. By 1939, the percentage had decreased even further. Only nineteen of forty-six students pursued farm-waste studies. The only bulletin during this decade was O. R. Sweeney and L. K. Arnold, "Plastics from Agricultural Materials," Iowa Engineering Experiment Station Bulletin #154 (Ames, 1942). Also in 1938, the chemical engineering department no longer could claim the greatest number of graduate students within the division. See Iowa State Board of Education, Biennial Report, 1938, 314.

35. Kent Knutson, "Fibers for Wars," *Iowa Engineer* 44 (1944), 117; Roger Williams, "Profit in Floss," *Iowa Engineer* 45 (1945), 177; Boris Berkman to Sweeney, 23 January 1945, and 10 April 1945; Sweeney to W. J. Schlick, 3 November 1944; Sweeney to Johart Larwill, 27 June 1946, all in Sweeney Papers; Iowa State Board of Education, Biennial Report, 1942, 20.

36. For Sweeney's and Arnold's work after the war, see, for example, O. R. Sweeney and L. K. Arnold, "Moisture Relations in the Manufacture and Use of Insulating Board," Iowa Engineering Experiment Station Bulletin #163 (Ames, 1948); O. R. Sweeney, L. K. Arnold, and E. G. Hollowell, "Extraction of Soybean Oil by Trichloroethylene," Iowa Engineering Experiment Station Bulletin #165 (Ames, 1949); and O. R. Sweeney and L. K. Arnold, "Furfural Utilization Studies," Iowa Engineering Experiment Station Bulletin #169 (Ames, 1950).

2   Changing Land-Grant Engineering from the Outside

Chemical Engineering, Accreditation, and the Land-Grant Colleges

Terry S. Reynolds

Introduction

In land-grant colleges for several decades after the passage of the Morrill Act in 1862, engineering education was generally inferior to that of the largely private schools that had pioneered academic engineering instruction in the United States. Land-grant schools had lower entrance requirements, fewer faculty, meager equipment.[1] Between 1890 and 1930, however, engineering instruction at land-grant colleges fully attained national norms. By the late 1920s around one-third of all engineering schools were land-grant schools, and half of total engineering student enrollment was in these schools.[2] A clear indication that land-grant engineering education had achieved equity was the important role land-grant schools played in two major developments in early-twentieth-century American engineering education: the emergence of chemical engineering programs and the development of engineering accreditation.

Accreditation is the voluntary acceptance of inspection based on standards set by external, non-governmental agencies as a means of credentialing education programs. It is a fact of life in American higher education today. Colleges and universities regularly submit to review by accrediting agencies, despite the immense investment in time and effort involved in preparing for reviews, because they view the alternatives—educational chaos or tight government oversight—as less acceptable. Accreditation is uniquely American. In most of the world control over higher education is vested in ministries of education with broad powers over admission standards, curriculum, and teaching credentials.[3] Reflecting early distrust of federal control over education, the American Constitution left education decentralized. Accreditation ultimately became the American alternative to centralized, government oversight.[4]

Accreditation first emerged shortly before 1900 for high schools, a process often directed by state universities. Shortly after 1900 professional schools in law, medicine, and dentistry accepted outside oversight, usually directed by professional associations, and by 1920 accreditation had been embraced by a variety of other fields, including business, landscape architecture, and nursing.[5]

Engineering, however, was slow to respond, accepting accreditation only in the mid- to late 1930s. Within engineering, the subdiscipline that pioneered the process was, ironically, chemical engineering, then the youngest and least firmly established of the major engineering disciplines. Its leading professional society—the American Institute of Chemical Engineers (AIChE)—began formal accreditation of chemical engineering programs a full decade earlier than the engineering profession generally.

This essay will focus on four issues: (1) why accreditation emerged late in engineering, (2) why chemical engineering led much older and larger engineering disciplines into accreditation, (3) what influence chemical engineering's early accreditation program had on broad-based engineering accreditation when it emerged, and (4) what role land-grant schools played in these developments.

Engineering's Late Start

Compared to other major professions, engineering entered the world of accreditation late. In part this was because engineering enrollments grew rapidly between 1890 and 1920, encouraging engineering educators to focus primarily on staying ahead of student demand for their programs.[6] Quality issues took a back seat.

But concern for quantity over quality was not the only reason that engineering was slow to embrace accreditation. The long historical disjunction between engineering education and engineering practice in America was more critical. American engineering education grew up autonomously.[7] Educational programs in engineering generally emerged before a strong professional consciousness developed. In civil engineering, for example, educational programs emerged in the 1820s and 1830s,[8] but a viable national professional society in the field—the American Society of Civil Engineers (ASCE)—did not solidify until 1867. The earliest mechanical engineering academic programs emerged in the 1850s and expanded rapidly in the 1870s as many of the new land-grant colleges interpreted their "mechanic arts" charge to mean "mechanical engineering." The American Society of Mechanical Engineers (ASME) emerged only in 1880.

Even after their founding, professional engineering societies cared little about academic education because engineering emerged in America as an open profession, entered initially through on-the-job experience much more frequently than through college credentialing.[9] Practically trained engineers early dominated professional engineering societies, and they typically had little interest in formal education.[10] Engineering professors played a minimal role in the founding or the early operations of professional engineering societies.[11] In fact, educators founded their own professional society, the Society for the Promotion of Engineering Education (SPEE) in 1893, in part because existing engineering groups had little interest in engineering education.[12]

As a result of this historic disjunction, engineering colleges developed a tradition of almost total freedom from external, professional control.[13] In 1926, in a private memorandum to the board that oversaw his monumental study of engineering education, William Wickenden noted: "There is no tradition more time-honored among the engineering colleges than that of freedom from external control." "It seems clear," he added, "that the engineering colleges will be unwilling to enter into any relationship among themselves and with any extra-academic bodies which will lessen appreciably their scope for individuality and initiative."[14] This long tradition of freedom from external control probably best explains the slowness of engineering education to embrace accreditation by external agencies.[15]

The Chemical Engineering Exception

The lack of interest by the major disciplinary engineering groups in establishing and enforcing standards left the pioneering role in accreditation to the youngest of the major disciplinary engineering societies: the American Institute of Chemical Engineers (AIChE).

The pattern by which chemical engineering emerged as a profession offered no hint that it would take a lead role in engineering accreditation. It seemed to follow the pattern of other engineering disciplines: educational programs preceded the development of a professional society, which, when it emerged, was dominated by practitioners who had some disdain for academic education.

MIT, Massachusetts's land-grant school for the mechanic arts, is widely recognized as inaugurating the first program explicitly labeled chemical engineering: its Course X of 1888–1889 combined mechanical engineering with industrial applications of chemistry. Other chemical engineering programs followed quickly. According to W. K. Lewis, one of the key figures in early chemical engineering at MIT, when the demand developed for chemical engineers "it was met primarily by the initiative of land-grant institutions."[16] While exaggerated, Lewis's statement contains an element of truth. Between 1888 and 1905 programs in chemical engineering emerged at a number of other land-grant schools, including at the University of Illinois in 1894, Ohio State University and the University of Arkansas in 1902, and Louisiana State University and the University of Wisconsin around 1905.[17]

Land-grant schools offered particularly favorable grounds for the emergence of chemical engineering, since, in their initial permutations, chemical engineering programs were largely combinations of chemistry and mechanical engineering, and both of these areas had early established very strong footholds in land-grant schools.

Chemistry had a strong position because promoters of agricultural education believed chemistry had a major role to play in the improvement of agriculture. Practically all land-grant schools secured chemists for their faculty very early,
and most agricultural curricula required at least one course in agricultural chemistry.[18] The emergence of agricultural experiment stations in the 1880s and 1890s
further expanded the strong position of chemistry in land-grant institutions.[19]

Mechanical engineering also quickly established a foothold in land-grant institutions, largely because most land-grant schools by 1880 interpreted the term
"mechanic arts" in the Morrill Act to mean mechanical engineering.[20] By 1900 engineering had become the dominant field on most land-grant campuses. Between
1889 and 1922, for example, 43 percent of the graduates of land-grant universities
came from engineering (versus 18 percent from agriculture), and within engineering, mechanical engineering was typically the largest unit.[21]

The land-grant colleges' tradition of providing practical service to the states
in which they were located further contributed to the emergence of chemical engineering programs.[22] For example, several of the key figures in the emergence of
chemical engineering at the University of Wisconsin emphasized its importance to
the development of Wisconsin's natural resources, such as its beet sugar industry.[23]
At Iowa State, the emphasis was slightly different but in the same tradition: the development of practical uses for agricultural waste products.[24] By 1930 at least
thirty-two land-grant colleges offered programs in chemical engineering.[25]

While the earliest academic programs in chemical engineering emerged in
the late 1880s, the field did not began fully to develop professional consciousness
until the formation of the AIChE in 1908, fully twenty years after MIT's initial offering. In keeping with the pattern of other engineering fields, the AIChE was
founded and initially dominated by practitioners, not academics. In fact, one
prominent New York consulting engineer at an early AIChE meeting declared that
academic training was "to a large extent valueless" and that the academic habit of
mind was "distinctly dangerous in technical work."[26] In its early years, the AIChE
frequently denied membership to academic chemical engineers due to their lack
of practical engineering experience or because, at the time they applied for admission, they could no longer be considered "practicing" chemical engineers (a requirement for membership).[27] In 1930 AIChE membership included only seventeen land-grant chemical engineering educators.[28]

Although chemical engineering seemed to fit the profile of other disciplinary groups in American engineering in the way it emerged, it differed in one major respect. Despite dominance by practitioners, the discipline's leading professional society (AIChE) exhibited an early concern with education.

At the June 1908 organizational meeting of the AIChE, Charles McKenna,
who delivered the keynote address, called improving chemical engineering education the "noblest aim" before the new society.[29] Others apparently agreed. In December 1908, at its first regular meeting, the Institute formed the permanent Com-

mittee on Chemical Engineering Education to determine just what the education of a chemical engineer should include.[30] In December 1913 the Committee recommended that the AIChE undertake a detailed study of chemical engineering education.[31] The small size of the Institute and the onset of World War I delayed implementation of the recommendation.

In 1919, with the war over, the AIChE renewed its concern with education. In December 1919 the Institute asked its Committee on Chemical Engineering Education to recommend specific steps the Institute could take to improve academic graduates. By late 1920 the Committee, now chaired by Arthur D. Little, had identified 77 institutions offering chemical engineering degrees and had begun to systematically gather, tabulate, and analyze data from these.[32] As might be expected, a large number of the chemical engineering programs were at land-grant universities: nearly 45 percent of the total. In 1921–22 the Committee continued its studies, making an interim report in 1921 and a final report in 1922.

The Committee found chemical engineering education in chaos, with a "wide divergence" in curricular requirements, course weighting, and course nomenclature. The Committee, however, reached consensus on what the core of a chemical engineering education should be and the steps that should be taken to reform offerings, including reduction in the multiplicity of required courses drawn from other engineering fields, avoidance of specialization by industry, standardization of course nomenclature and weighting, and the initiation of course work centered around the concept of "unit operations" (more below). Because no authority existed to implement these recommendations and the need for reform was "evident and urgent," the Committee's chair, Arthur Little, suggested that the AIChE sponsor a meeting with representatives from educational institutions to discuss the report.[33] The Institute accepted the report and recommendations enthusiastically. Ralph McKee, a member on the Institute's Board, noted that the Committee had "written a prescription," and that the Institute's duty was to see that it was filled and given to the patients.[34]

Not unnaturally, in view of the large number of chemical engineering programs at land-grant universities, the faculty and graduates of land-grant schools played an important role in the AIChE's Committee on Chemical Engineering Education. The driving force behind its systematic study of chemical engineering education was Arthur D. Little, a graduate of MIT, Massachusetts's land-grant school for the "mechanic arts," who continued to be closely affiliated with that school's chemical engineering program, even while running a private consulting firm. The Committee's chair preceding Little and vice-chair under Little was James Withrow of Ohio State University, another land-grant institution.

The AIChE's conference with educators, held at Brooklyn Polytechnic in May 1922, made only minor changes in the Little Committee's report. At its June 1922 meeting the Institute accepted the revised report and appointed a new Committee on Chemical Engineering Education consisting of five members from industry and five from education, instructing it to continue studying chemical engineering programs, attempt to persuade colleges to accept the conclusions of the Little report, and to publish, after a three year period, the names of institutions whose programs met the guidelines laid down in Little's report; in other words, it initiated a program of accreditation.[35]

The Committee worked at this task between 1922 and 1925, developing a classification for courses, setting up a tentative norm for comparison, and reviewing programs at a large number of schools. It established preliminary standards in areas such as entrance requirements, organizational autonomy, faculty qualifications, and physical facilities. H. C. Parmelee succeeded Little as chair of the Committee when the AIChE moved from study to implementation. Parmelee's Committee maintained strong representation from land-grant schools. Parmelee, editor of the field's leading trade publication (*Chemical and Metallurgical Engineering*), was a graduate of a land-grant school: the University of Nebraska. Three other members of the Committee also had associations with land-grant schools: W. K. Lewis (MIT), Arthur D. Little (MIT), and Samuel Parr (Illinois).

In June 1925 the Committee presented its work, including a list of fourteen schools regarded as meeting its standards, recommending that the AIChE establish a permanent committee to periodically review schools.[36] Of the fourteen schools approved, or accredited, by the AIChE, land-grant schools were represented in proportion to their overall numbers. Nearly 45 percent of the schools that Little's 1922 survey identified as having chemical engineering programs were land-grant schools; nearly 45 percent of the schools accredited (six of fourteen) were either land-grant schools or former land-grant schools.

The AIChE's Council accepted the Committee's report. In March 1927, following a questionnaire on whether the AIChE should remain involved in accreditation, the Council concluded that the Institute should continue.[37] So successful was the program, with schools clamoring to make the changes necessary to get on the AIChE's approved list, that in 1928 H. C. Parmelee asserted that the AIChE's educational program was "outstanding among the Institute's activities" in its first twenty-five years.[38]

Why Chemical Engineering?

In undertaking systematic accreditation and adopting it as a professional responsibility, the AIChE had undertaken a program unique among contemporary professional engineering societies. In 1922, when the AIChE plunged ahead with accreditation, the remainder of the engineering profession had barely begun to consider the prospects. Why was this? Why should the youngest and least well defined of the major engineering disciplines go where others had feared to tread? To answer these questions, we must step backwards and look at the professional concerns of chemists involved in chemical manufacturing around 1900, for it was their real or perceived vulnerabilities that best explain what drove chemical engineering into the uncharted territory of accreditation.

The emergence of chemical engineering was not the inevitable result of the expansion of the chemical industries. In the late nineteenth century, the world's leading chemical power was Germany. In Germany, research chemists working in cooperation with or assisted by engineers, physicists, and metallurgists carried out the engineering work involved in designing, scaling up, and operating chemical plants.[39] One of the deans of the German chemical industry, Carl Duisberg, argued in 1896 that the German tradition of division of labor between a chemist untrained in engineering and an engineer untrained in chemistry was fully within the American tradition of division of labor and "absolutely necessary."[40] The leading American society representing chemists—the American Chemical Society—agreed and actively opposed the emergence of chemical engineers as a distinct professional grouping.[41]

Moreover, in the United States, mechanical engineers were, by 1900, in the process of establishing themselves not only as partners with chemists in the design and operation of chemical plants, but in many cases as the dominant partners—the reverse of the German scheme, where chemists dominated. Chemists engaged in manufacturing in America viewed this trend with trepidation. In 1902 C. F. Burgess, later a founder of the chemical engineering program at the University of Wisconsin, complained that the heads of the largest chemical industries were often engineers with mechanical or electrical training, not chemical training.[42] In 1911 another chemical engineer grumbled that the design of chemical plants was "too many times" left to architects, civil, or mechanical engineers who did not know the requirements to operate the plant.[43]

That the engineer was coming to play a greater role than the chemist in the American chemical industry—in contrast to the German—is not surprising. The German industry focused on producing a large number of chemically complex substances in relatively small volumes, such as synthetic dyes and pharmaceuticals, an approach that gave dominance to chemical considerations. The early American chemical industry, on the other hand, focused on producing very large quantities of a few chemically simple substances, such as sulfuric acid, nitric acid, bleach, soda, and superphosphate fertilizers, an emphasis on quantity over quality that gave dominance to mechanical, or engineering, considerations.[44]

The relatively low public image of the chemist in the Unites States made matters worse. In Germany the chemist, particularly the research chemistry, was viewed with awe and respect.[45] In the United States this was not the case. Most American chemists were involved not in research but in routine chemical analysis, often confused with druggists, and viewed as having little capacity to manage industrial scale operations.[46] This perception undermined the elite chemists who were involved in plant design and operation and who hoped to expand their role in the American chemical industry.

It was a small group of elite chemists—those already involved in supervision of manufacturing and in chemical plant design—who embraced the concept of "chemical engineer" and founded the AIChE. The idea of a "chemical engineer" as a unique professional capable of handling both chemical and engineering problems can be seen as part of a two-prong strategy on the part of these chemists. First, the term "engineer" in "chemical engineer" enabled them to differentiate themselves from the low-status analytical chemists who also worked in industrial settings but who were not involved in design or supervision of labor. Second, the concept of the chemical engineer as a special professional with *both* engineering *and* chemical skills offered a way for chemically-trained professionals to reclaim dominance from mechanical and electrical engineers in the design and operation of plants in the chemical industry.

The path to the creation of the new engineering profession was not an easy one. The question of "What is a chemical engineer?" said John C. Olson, the AIChE's first secretary, "bobbed up incessantly" at early meetings of the Institute.[47] For years considerable disagreement persisted over whether chemical engineers should consider themselves primarily chemists or primarily engineers. One definition of chemical engineering in circulation suggested that a chemical engineer was one who talked "engineering in the company of chemists, chemistry in the company of engineers, and politics when in the company of both simultaneously."[48] Thus, some felt that the new field was nothing more than a hybrid, a wholly artificial construct with no real contribution to make over and above that made by chemists and mechanical engineers separately.

As a result, chemical engineers had considerable difficulty in securing recognition from other engineers or from chemists. When the Society for the Promotion of Engineering Education (SPEE) initiated its first major investigation of engineering education in conjunction with the Carnegie Foundation (the Mann investigation) around 1910, the AIChE was not represented. Charles R. Mann, the study's principal investigator, regarded the American Chemical Society as the spokesman for chemical engineering education.[49] When the Naval Consulting Board was organized in 1915, authorities invited the societies representing civil, mechanical, electrical, and mining engineering to participate, as well as the American Chemical Society and even the American Electrochemical Society, but not the American Institute of Chemical Engineers.[50] It would be easy to multiply examples.[51] Responding to these snubs Harry O. Chute, a New York–based consulting chemical engineer, lamented in 1921: "we are not even able to convince other engineers that we are engineers."[52]

Academic programs in chemical engineering reflected the confusion the field had in defining its intellectual boundaries. In 1910 the president of the Polytechnic Institute of Brooklyn, Frederick W. Atkinson, noted that training in chemical engineering was "chaotic" because technical schools could not agree on the most suitable course of education. It may safely be said, he added, "that no two institutions have from among an almost bewildering variety of engineering subjects even made even approximately the same selections."[53]

Thus, the AIChE's founders faced problems not really encountered by earlier engineering fields, or encountered by them at a much lower scale of magnitude: lack of recognizable intellectual boundaries, the need to differentiate chemical engineering from chemistry, difficulties in securing recognition from other engineering disciplines, and uncertainty about what the education of a chemical engineer should include. These related problems best explain chemical engineering's atypical early interest in education, for chemical engineers recognized that it was in the educational arena that intellectual boundaries were defined, defended, and passed on to coming generations.

The primary need was for a conception of chemical engineering that made it more than an artificial and sterile hybrid of chemistry and mechanical engineering. The person who popularized that conception—"unit operations"—was Arthur D. Little, a graduate of MIT. In 1915, while serving as a member of MIT's governing body, Little chaired a committee charged with reviewing MIT's chemical engineering program. In his report to the MIT Corporation he maintained that chemical engineering education should not imitate industrial chemistry by tracing chemical processes in individual industries. Instead, he argued, it should focus on those "unit operations" that cut across industrial lines, that were universal to all chemical processes, such as pulverizing, dyeing, roasting, crystallizing, filtering, and evaporation.[54] In 1922, he reiterated this point, asserting:

> Chemical engineering as a science, as distinguished from the aggregate number of subjects comprised in courses of that name, is not a composite of chemistry and mechanical and civil engineering, but a science of itself, the basis of which is those unit operations which in their proper sequence and coordination constitute a chemical process as conducted on the industrial scale. . . . Their treatment in the quantitative way with proper exposition of the laws controlling them and of the materials and equipment concerned in them is the province of chemical engineering. It is this selective emphasis on the unit operations themselves in their quantitative aspects that differentiates chemical engineering from industrial chemistry . . .[55]

Little's proposed use of "unit operations" as a way to delineate the intellectual boundaries of chemical engineering was quickly embraced by the struggling discipline.

Unit operations provided chemical engineering with an intellectual realm of its own and, simultaneously, provided the field with an organizing principle for its educational programs: the analytical study of unit operations, which occurred across a variety of industries, instead of the descriptive study of particular industries. The AIChE's long interest in chemical engineering education ensured that unit operations' implications for clearing up the chaotic nature of educational programs in chemical engineering would be quickly recognized. In discussing a paper in December 1917, for example, Launcelot Andrews, with the Victor Chemical Works, noted that current chemical engineering education was based on "an entirely incorrect system," where subject matter was classified according to industry, not on the operations common to a variety of industries. He argued that instruction, instead, "might be almost entirely" focused on the study of those elements fundamental to all large-scale chemical operations. The comment was greeted with applause.[56]

By appointing Little to chair its Committee on Chemical Engineering Education in 1919, the Institute acknowledged his role in clarifying and publicizing the "unit operations" concept. Little's recommendation in his 1922 report to the AIChE that the chaotic nature of chemical engineering education be remedied by organizing it around unit operations was not a big surprise. At the conference held by the AIChE with educators in 1922, both educators and employers of chemical engineers were reportedly "completely in accord" on a number of points, including the need to organize chemical engineering education around unit operations.[57] In a discussion on chemical engineering education, 1925 AIChE president Charles Reese commented: "What is a chemical engineer? I think the thing is crystallizing now . . ."[58]

Accreditation provided a mechanism to enable the AIChE to ensure that academic programs embraced the newly developed definition of chemical engineering, which centered around unit operations, and put in place appropriate mechanisms to separate chemical engineering intellectually and institutionally from chemistry. Thus, the AIChE's accreditation guidelines included not only organization of curricula around unit operations, but also moving chemical engineering programs institutionally from chemistry departments to engineering colleges and insistence that chemical engineering courses be taught by chemical engineers, not chemists. The AIChE was surprised by the efforts schools were willing to make in order to get on its list of approved schools.[59] H. C. Parmelee pointed out in the early 1930s that the AIChE's educational activities had led to "widespread acceptance" of a uniform, basic curriculum in the field.[60]

The Expansion of Accreditation

The AIChE's success in initiating accreditation between 1922 and 1925 did not lead directly to the broader adoption of accreditation by the engineering profession, which occurred in the period 1932–1936. The broader engineering profession's involvement in accreditation seems, instead, to have emerged out of the comprehensive study of engineering education undertaken by the Society for the Promotion of Engineering Education, beginning in 1923 and completed around 1930, a study directed by William Wickenden.

By 1920 some engineering educators had become concerned that engineering education had fallen relatively behind other fields of professional education.[61]

For instance, C. L. Mees, former president of Rose Polytechnic Institute, noted that whereas engineering education had been more professional than medical or legal education in 1900, by 1920 the reverse was the situation. He added that it was "very doubtful" if engineering colleges "should be called professional colleges at all."[62]

Hence when Charles Scott, as president of SPEE in 1922 and 1923, proposed a massive new study to determine how to improve and enrich engineering education and in 1923 secured funds ($108,000) for such a study from the Carnegie Foundation, SPEE embraced the project. Scott and his colleagues persuaded William Wickenden, then a vice president for educational programs at AT&T, to direct the inquiry under the supervision of a five-member board.[63] Land-grant schools were initially not well represented on the oversight board, although Wickenden, who had taught at the University of Wisconsin and MIT before taking a post at AT&T, certainly had a land-grant school background. Even though land-grant schools by the 1920s accounted for around 50 percent of all engineering students, only one of the five board members came from a land-grant school: Dugald Jackson of MIT. This prompted some internal correspondence focused on getting a representative of a western land-grant school on the Board.[64] That did not happen immediately, but of the eight engineering educators who served on the Board between 1923 and 1930, fully half ultimately came from land-grant schools.

The Wickenden-directed SPEE study was the most extensive ever undertaken of engineering education, making an effort to involve every American engineering school, representatives of American industry, and the major professional engineering societies in all stages from data gathering to analysis to implementation of recommendations. The study looked at every aspect of engineering education from the preparation of entering students to curricular organization to faculty salaries to career performance after leaving college.

The first year or so of the investigation focused on data gathering. But when Wickenden turned to data analysis, he faced the issue of how recommendations drawn from the massive volumes of data accumulated could be implemented. The implementation issue led him, naturally, to consider the possibility of giving engineering societies a role in engineering education. In a private memorandum to the Board of Investigation and Coordination that supervised the study, dated March 4, 1926, Wickenden noted that the most striking development in higher education in the previous quarter century had been the increased control of professional schools by national professional organizations. No study relating to engineering education, he asserted, could neglect this.[65] In both private and public reports on his work from this time forward, and in his final recommendations, Wickenden argued for more standardization of engineering curricula, for the use of some form of accreditation to accomplish this, and for greater cooperation between educational institutions and professional engineering societies.[66] In a lecture delivered to the American Society of Mechanical Engineers in December 1927, he publicly promoted the idea of engineering society involvement in standardizing and accrediting engineering education.[67]

In 1930, as the Wickenden study wound to a close, the SPEE's Council appointed a committee, headed by Wickenden, to consider "classification" (i.e., accreditation) of engineering schools. The committee submitted a report to SPEE in June 1931 in which it reported finding sentiment for establishing and enforcing standards under the aegis of SPEE and that the officers of certain national societies had "intimated" that they would welcome the SPEE initiative in the area.[68] Not surprisingly, since enrollments in engineering in land-grant schools approached 50 percent of total engineering enrollments, this SPEE Committee, like Wickenden's Board of Investigation and Cooperation, had substantial representation from land-grant colleges, including Robert Sackett of Penn State College, and William Wickenden, who, as noted previously, had taught at the land-grant University of Wisconsin earlier in his career. But SPEE hesitated to act alone due to the magnitude of the project and its limited resources.

The spread of engineering licensing laws, however, spurred both SPEE and the disciplinary professional engineering societies to further action. Engineering societies generally opposed licensing laws for a variety of reasons including ideological distrust of the involvement of government in credentialing engineering practitioners and the desire of senior engineers to keep the cost of engineering employees down by insuring no restrictions (such as licensing requirements) on the supply. Thus the leaders of the professional engineering societies saw development of some form of certification—other than licensing—as an alternative to licensing laws. Training in engineering schools was seen as one element of this alternative certification. SPEE and the engineering education community were concerned that most state licensing laws tended to specify graduation from a "recognized" college, and since there was no national-level list of "recognized" engineering colleges, some state licensing boards had begun creating their own, often mutually contradictory, lists.[69] Moreover, the national group representing state licensing boards—the National Council of State Boards of Engineering Examiners (NCSBEE)—had appointed a committee in 1931 to deal with the "recognized" college issue on a national basis.[70] With clear mutual interests, the most activist of the "founder" societies, the American Society of Mechanical Engineers (ASME) and the SPEE, issued an invitation to the other major societies to participate in a conference on certification into the engineering profession. The invitation included the AIChE and the NCSBEE as well.[71]

At this conference in April 1932, the delegates recommended a common plan of action. This plan called for creation of a new, joint body—the Engineers' Council for Professional Development (ECPD)—one of whose functions was to be to "formulate criteria for colleges of engineering, which will insure their graduates a sound educational background for practicing the engineering profession." Although all participating societies did not formally endorse the ECPD as the official agency for accrediting educational programs until early 1935, the ECPD's Council appointed a "Committee on Engineering Schools" in November 1932 to prepare criteria and processes for accrediting engineering schools.[72] The committee had solid representation from land-grant schools. The committee's chair, Karl Compton, was from MIT. The seven-man committee also included four other faculty or administrators from land-grant colleges: A. A. Potter from Purdue, G. M. Butler from the University of Arizona, Ivan Crawford from the University of Idaho, and P. H. Daggett from Rutgers. A representative from a land-grant college chaired the other primary educationally oriented committee of the ECPD as well: Dean R. L. Sackett of Penn State College chaired the Committee on Student Selection and guidance. ECPD's first list of accredited schools appeared in 1936.

The AIChE's Role

What role did the AIChE's accreditation program play in the adoption of accreditation by the engineering profession as a whole? As noted above, the primary drive for accreditation emerged out of Wickenden's study, with the timing influenced by the spread of state licensing laws, which, in turn, were in part the result of engineering unemployment and declining salaries in the early years of the Great Depression.

The AIChE's pioneering accreditation program, nonetheless, seems to have had an influence on the emergence of engineering accreditation in America at two distinct points: in late 1925 or early 1926, when Wickenden first began seriously considering how to implement the results of his study; and in 1932, when the ECPD began laying the foundation for profession-wide accreditation.

It seems likely that William Wickenden, when he began his study of engineering education in October 1923, was unaware of the AIChE's work. In a private memorandum to the Board of Investigation in March 1926, he commented that the AIChE's program had "not attracted wide attention," which at least suggests that he had not heard of it before he began work.[73] This would not be surprising. The AIChE in 1923 had not yet published its first "approved" list, and while the other "founder" societies had memberships either approaching or well in excess of ten thousand by the mid-1920s, AIChE membership was only around five hundred. It was a small society, easily overlooked.

Apparently, Wickenden became aware of the AIChE's efforts no later than early 1925. At the June 1925 AIChE meeting, Little reported that the Society for the Promotion of Engineering Education had displayed "a lively interest" in the work of the AIChE's Committee on Chemical Engineering Education. He also reported that Wickenden had informed him that chemical engineering was "better organized today than any of the other branches [of engineering] from the point of view of the definite objective that it has in mind, and the product that it turns out to meet certain qualifications."[74] Despite these conversations, Wickenden, in a private report to his Board of Investigation on October 16, 1925, noted contact with the AIChE but indicated wariness about involving the Institute and other minor societies more directly in the Board's work.[75]

In December 1925 in a series of memos to the Board of Investigation, Wickenden began to focus on the issue of standardization because he was concerned about how to ensure that the recommendations likely to emerge from his study would be implemented. Standardization blended naturally into accreditation, since accreditation was clearly a means of enforcing standards. In a March 1926 memo he specifically noted that the AIChE was the exception among engineering groups in having and enforcing standards through an accreditation program, in spite of adverse criticism from institutions given an unfavorable classification.[76] The following month Wickenden attended an AIChE conference on chemical engineering education where the AIChE discussed the standards used to create its first accredited list of fourteen programs.[77] At about the same time he recommended to the Board of Investigation that it ask the AIChE to appoint a representative to it.[78]

It appears, then, that in a period of six months between October 1925 and March 1926, Wickenden went from wanting to keep the AIChE at arm's length to consulting with its educational leaders, attending its conferences, and inviting closer cooperation. I conclude from this sequence of events that as Wickenden turned from data analysis to working out how recommendations for improvement could be implemented, he began to see accreditation as a key tool. He could embrace that tool, in part, because the AIChE's experience had demonstrated that accreditation by professional societies could be used effectively to enforce standards on engineering schools, in spite of their long tradition of independence from external control. This recognition prompted his sudden interest in the professional group that he had at first wanted to keep at arm's length and led to his eventual conviction that professional engineering societies in America should become more involved in educational activities through accreditation.

In his final published report in 1930, Wickenden noted that chemical engineers stood out "conspicuously" among the various groups of engineers in favoring a greater role by national engineering societies in engineering education and in favor of a system to recognize colleges meeting approved minimum requirements. Wickenden added: "No doubt the experience of the A.I.Ch.E. in scrutinizing curricula in chemical engineering and in publishing a list of those approved by its educational committee was a considerable factor in the more positive and favorable expression than that given by other groups of engineers."[79]

The second point at which the AIChE influenced the emergence of engineering accreditation in America came following the formation of the ECPD in 1932. Early ECPD records are not publicly available, but it appears that the form ECPD-directed accreditation took was strongly influenced by AIChE. Any program of accreditation had to take into account the AIChE's already existing and successful program. This existing program enabled the AIChE to negotiate a unique position in the ECPD reporting format, a position that continues to this day. The AIChE's participation in the ECPD takes the form of a separate AIChE Committee on Education and Accreditation, which handles chemical engineering curricula, a departure from regular procedure made "in order that advantage might be taken of the considerable experience gained by AIChE in the course of its well established accrediting program."[80] The procedure adopted called for the ECPD to automatically accredit chemical engineering curricula on the existing AIChE list. Moreover, AIChE inspectors on ECPD teams would report first, not to the ECPD, but to the AIChE's Committee of Education, and the AIChE's Committee would then transmit a formal report and recommendation to the ECPD.[81] An ECPD-appointed committee would handle all other fields jointly.

The accreditation procedures adopted by ECPD undoubtedly drew from the experiences of a large number of professional groups. But the AIChE's procedures probably provided the immediate model. The chair of the Committee on Engineering Colleges of the ECPD, which established the ECPD's accreditation procedures and standards, noted in 1933 that while existing accreditation focused on institutions and was rather loose, in chemical engineering, accrediting was done "in accordance with quite definitely stated procedures and specifications." He added that, furthermore, the AIChE "keeps tab on institutions it has accredited and revisits them at intervals."[82] Not surprisingly, the ECPD's procedures resembled the AIChE's in a number of particulars. First, the ECPD, like the AIChE, first gathered information by questionnaires, followed by visits of inspection and periodic revisits. Second, the ECPD, like the AIChE, made no attempt to classify schools in groups according to rank. Programs either satisfied the minimum standards or they did not. Only lists of those who met the standards were published.[83] Finally, ECPD accreditation was by engineering curriculum (i.e., chemical engineering, mechanical engineering, civil engineering, etc.), not by engineering colleges as a whole. The ECPD's decision to accredit by curricula rather than by school was likely forced by the prior existence of the AIChE's program. Certainly, Wickenden's early discussions of accreditation focused on schools rather than individual curricula, and the NCSBEE certainly planned on accrediting entire schools, not individual curricula.[84]

The AIChE's leaders at the time were certainly convinced that their program had influenced the guidelines adopted by the ECPD. H. C. Parmelee, chair of the AIChE's Committee on Chemical Engineering Education, for instance, noted in 1932, as ECPD was in the process of formation, that there were "indications that its [AIChE's] principles and methods are likely to be adopted by other engineering groups."[85] Non-chemical engineers have also occasionally acknowledged the AIChE's influence on ECPD. For example, in 1981 Lee J. Walker, president of the ECPD's successor organization, the Accreditation Board for Engineering & Technology (ABET), told the AIChE's Council that the AIChE "was the model through which ECPD developed."[86]

Conclusion

The role of land-grant schools in the emergence of chemical engineering and engineering accreditation reflects their achievement of equity in engineering education with private and non-land-grant public universities by the turn of the century and shortly thereafter. The strong position of chemistry and mechanical engineering in land-grant schools—the result of their emphasis on agriculture, the mechanic arts, and practical education—placed land-grant schools at the forefront in the emergence of chemical engineering programs. Although land-grant schools provided only about a third of the nation's engineering programs in the 1920s, they provided nearly 45 percent of its chemical engineering programs. Engineers associated with land-grant schools usually chaired the various committees of the American Institute of Chemical Engineers that directed its pioneering efforts in engineering accreditation, and the organization's first list of accredited schools contained land-grant schools in proportion to their overall numbers among chemical engineering programs.

In organizing the committee to guide Wickenden's monumental study of engineering education, SPEE officials worried about appropriate representation for land-grant schools, and, ultimately, representatives from land-grant schools composed about 50 percent of all those who served on the Board. Finally, as we have seen, land-grant schools had appropriate representation on the key committees of the Engineers' Council for Professional Development, the organization that initiated accreditation of engineering more widely in the early 1930s.

The prominent role that representatives of land-grant schools played in the emergence of chemical engineering and in the development of engineering accreditation provides clear evidence that land-grant engineering programs had by the period 1890–1930 achieved equity with engineering programs at endowed polytechnic institutes, endowed universities, and non-land-grant state universities. By 1930, even if land-grant institutions had not really developed a distinctive form of engineering education, they had evolved to the point that their programs were fully representative of the general tendencies in American engineering education.[87]

Notes

Research for this study was partially conducted under the auspices of a grant from the National Science Foundation (DIR-892-1936).

1. U.S. Department of the Interior, *Survey of Land-Grant Colleges and Universities* [*Bulletin*, 1930, no. 9, vol. 1, pt. X: Engineering] (Washington: Govt. Printing Office, 1930), p. 800; James G. McGivern, *First Hundred Years of Engineering Education in the United States (1807–1907)* (Spokane: Gonzaga University Press, 1960), pp. 95–97.

2. U.S. Dept. of Interior, *Survey of Land-Grant Colleges*, 1930, pp. 791, 844.

3. Lloyd E. Blauch, ed., *Accreditation in Higher Education* (Washington: U.S. Department of Health, Education, and Welfare, 1959), p. 3; William K. LeBold, Warren E. Howland, and Joseph L. McCarthy, "Accreditation Related to Engineering and Graduate Education: A Historical Review," *Journal of Engineering Education* 55, no. 6 (Feb. 1965): 175; and Frank G. Dickey, "The Accrediting Association at a Time of Change, Crisis, and Challenge," *Journal of Engineering Education* 58 (Dec. 1967): 287–288.

4. Hugh Hawkins, *Banding Together: The Rise of National Associations in American Higher Education, 1887–1950* (Baltimore: Johns Hopkins University Press, 1992), esp. pp. xi–xii, 78. Of course, there was no objection to federal *support* of education. Thus some universities, banded together to use united action as a path to increased federal support, such as the Land-Grant College Association (ibid., pp. 4–10). See also Blauch, *Accreditation*, pp. 9–10.

5. See LeBold et al., "Accreditation," pp. 175–181, for a general review of the history of accreditation; for engineering accreditation specifically, see pp. 181–186. See also William Wickenden and Adelaide Dick, "Professional Organizations and Professional Schools," *Proceedings of the Society for the Promotion of Engineering Education* [Hereafter *SPEE Proc*] 32 (1924): 234–244, and Blauch, *Accreditation*, pp. 224–228.

6. Hammond, "Promotion of Engineering Education in the Past Forty Years," *SPEE Proc* 41 (1933): 46, estimated that enrollment in engineering jumped from around 12,000 in 1893 to around 65,000 by 1933. Data provided in Charles R. Mann's *A Study of Engineering Education (Bulletin no. 11)* (New York: Carnegie Foundation for the Advancement of Teaching, 1918), p. 7, indicates that the average number of engineers graduating per year was less than 400 in the 1880s, but had reached over 3400 by the period 1911–1915.

7. Society for the Promotion of Engineering Education, *Report of the Investigation of Engineering Education* [hereafter *Wickenden Report*], vol. 1 (Pittsburgh: SPEE, 1930), p. 823. See also Wickenden, "Memorandum for the Board of Investigation and Cooperation," Feb. 18, 1928 [Wickenden Committee Archival Records, American Society for Engineering Education, Washington, DC].

8. Terry S. Reynolds, "The Education of Engineers in America before the Morrill Act of 1862," *History of Education Quarterly* 32 (Winter 1992): 459–482, points out that a much larger number of schools offered engineering training in the ante-bellum era than is usually supposed, with a number of offerings coming as early as the 1820s and 1830s.

9. See, for example, Daniel H. Calhoun, *The American Civil Engineer: Origins and Conflict* (Cambridge, Mass.: The Technology Press, 1960), pp. 24–53.

10. *Wickenden Report*, vol. 1, pp. 631, 650, 823, 1000, 1001. According to Wickenden (p. 631), in 1890 less than a quarter of the members of national engineering societies were college graduates. See also Wickenden and Dick, "Professional Organizations," pp. 231–232. Charles F. Scott, "What New Trends in Engineering Education?" *SPEE Proc* 46 (1938): 693, noted that in its first quarter century ASME had only two or three presidents with college degrees. He also noted that

engineering societies had "not vitally concerned themselves with education" (ibid.).

11. Wickenden and Dick, "Professional Organizations," pp. 224f. Peter Lundgreen, "Engineering Education in Europe and the U.S.A., 1750–1930," *Annals of Science* 47 (1990): pp. 72–73, notes the tendency of American professional engineering societies to stay aloof from educational matters. See also Wickenden, "Memorandum for Board of Investigation and Coordination," March 4, 1926 [Wickenden Committee Archival Records], where he notes that professional bodies in engineering had done "little or nothing" to foster technical education and had been "too detached." The "Report of the Committee on the Advisability and Practicability of Classifying the Engineering Colleges," *SPEE Proc* 39 (1931): 207, notes that leading engineering societies until the 1920s "held rather rigidly to the attitude that their function was merely to foster and disseminate ... technical matters." See also ibid., p. 209, where the authors note that lack of concern for education was in part the result of the "extreme individualism of the self-trained engineers of the pioneer period."

12. "Report of the Committee on Advisability," p. 200.

13. Wickenden and Dick, "Professional Organizations," pp. 231–233. According to Dugald C. Jackson, *Present Status and Trends of Engineering Education in the United States* (New York: ECPD, 1939), p. 3, "Up until that time [the 1920s], it had not been customary for the great national societies in the professional branches of engineering as a group to cooperate with the engineering schools ..." [A. M. Wellington], "The Ideal Engineering School, IV," *Engineering News* 29 (June 1, 1893): 513–514, complained that engineering professors had become too much "a class apart" from those who practiced engineering and criticized the circular that led to the formation of the Society for the Promotion of Engineering Education on that basis.

14. Wickenden, "Memorandum for the Board of Investigation and Coordination," March 4, 1926 [Wickenden Committee Archival Records], and Wickenden, "Memorandum for Board: Report to Counselors on National Engineering Societies," Feb. 18, 1928 [Wickenden Committee Archival Records]. Engineering professional societies did occasionally talk about the need to improve engineering education, but they generally believed that schools would not allow professional societies to dictate to them. See William H. Wisely, *The American Civil Engineer, 1852–1974* (New York: American Society of Civil Engineers, 1974), pp. 80–81.

15. Jackson, *Present Status,* pp. 20–21, also notes the independence of engineering faculties from each other.

16. W. K. Lewis, "Chemical Engineering," in U.S. Bureau of Education, *Land-Grant College Education, 1910–1920* [*Bulletin,* 1925, no. 5], Pt. IV: Engineering and Mechanic Arts (Washington: GPO, 1925), p. 21.

17. J. W. Westwater, "The Beginnings of Chemical Engineering in the USA," in William F. Further, ed., *History of Chemical Engineering* (Washington: American Chemical Society, 1980), pp. 142–145, 146–151. A number of non–land-grant schools also initiated programs early, including the University of Michi-

gan, the University of Pennsylvania, the University of Colorado, Tulane, and Tufts.

18. "Industrial Education in the United States," United States Senate, 47th Cong., 2nd Session, Ex. Doc. No. 25, December 27, 1882, pp. 19, 21, 24. See also, as a specific example, Waterman T. Hewett, *Cornell University: A History* (New York: The University Publishing Society, 1905), vol. 2, pp. 162–165.

19. Edward H. Beardsley, *The Rise of the American Chemical Profession, 1850–1900* (Gainesville: University of Florida Press, 1964), pp. 44, 53–55, and U.S. Bureau of Education, *Report of the Commissioner of Education*, v. 2 (Washington: GPO, 1901), pp. 1813–1826.

20. Martha M. Trescott, "Unit Operations in the Chemical Industry: An American Innovation in Modern Chemical Engineering," in William F. Furter, ed., *A Century of Chemical Engineering* (New York: Plenum Press, 1982), p. 6.

21. Mary Jean Bowan, "The Land-Grant Colleges and Universities in Human Resource Development," *Journal of Economic History* 22 (1962): 532. See also U.S. Department of Interior, *Survey of Land-Grant Colleges, 1930*, pp. 799–800. The Records of the Office of Assistant Commissioner, Division of Higher Education, U.S. Office of Education (RG 12), Office File of the Specialists in Land-Grant College Statistics, 1911–1926, Box. No. 1, Entry 41, File: San Francisco Exposition, 1915 (U.S. National Archives, Washington, DC), contains a chart of engineering enrollment at land-grant colleges and universities in 1903 and 1913. In both these years, enrollment in mechanical engineering at land-grant universities exceeded that of any other engineering field.

22. H. L. Plants and C. A. Arents, "The Evolution of Engineering in Land-Grant Institutions: History of Engineering Education in the Land-Grant Movement," *Proceedings of the American Association of Land-Grant Colleges and State Universities*, 75th Annual Convention-Centennial Convocation, vol. 2, Nov. 12–16, 1961, p. 88. See also *Proceedings of the 3rd Annual Meeting of the Land Grant College Engineering Association*, Washington, Nov. 10–13, 1914, pp. 9–10, 13.

23. John B. Johnson, "Some Unrecognized Functions of Our State Universities," *Journal of the Western Society of Engineers* 4 (1899): 376–387; and Magnus Swenson, "The Chemical Engineer," *Bulletin of the University of Wisconsin*, no. 39, Engineering Series, 2 (1900): 195–207.

24. Alan I Marcus and Erik Lokensgard, "The Chemical Engineers of Iowa State College: Transforming Agricultural Wastes and an Institution, 1920–1940," *Annals of Iowa* 48 (1986): 177–205.

25. U.S. Department of the Interior, *Survey of Land-Grant Colleges, 1930*, p. 829.

26. William Grosvenor in *AIChE Bulletin* 3 (Dec. 1910): 5.

27. Reynolds, "Defining Professional Boundaries: Chemical Engineering in the Early 20th Century," *Technology & Culture* 27 (Oct. 1986): 708; and Reynolds, *75 Years of Progress: A History of the American Institute of Chemical Engineers, 1908–1983* (New York: AIChE, 1983), p. 7. See also *Transactions of the American Institute of Chemical Engineers* [Hereafter *AIChE Trans*] 1 (1908): 21–22, for the membership requirements. Periodic discussions over the issue of mem-

bership requirements appeared frequently in *Bulletin of the American Institute of Chemical Engineers* [hereafter *AIChE Bulletin*]. See, for example, *AIChE Bulletin* 15 (1917): 6; 17 (1918): 21, and, especially, 20 (Dec. 1919), pp. 7, 13–14, 17–18. See also *AIChE Bulletin* 22 (Dec. 1920): 41. Academics made up less than 10% of society membership in 1909, and even that 10% gained admission mainly on the basis of their consulting work, not their academic positions.

28. U.S. Department of the Interior, *Survey of Land-Grant Colleges, 1930*, p. 818.

29. Charles F. McKenna, "The Justification of the American Institute of Chemical Engineers," *AIChE Trans* 1 (1908): 18.

30. *AIChE Trans* 1 (1908): 33. Committees often are little more than window dressing, but this Committee was the most dynamic element within the Institute over the next two decades. In 1910, for example, it led a discussion of chemical engineering education that was one of the central events of the annual Institute meeting, see *AIChE Bulletin* 4 (Oct. 1911): 9–21; "Report of Committee on Chemical Engineering Education," *AIChE Trans* 3 (1910): 122–152, and the papers immediately following its report.

31. *AIChE Bulletin* 8 (Dec. 1913): pp. 14–18, 44, 46; *AIChE Trans* 7 (1914): 18–25.

32. *AIChE Bulletin* 20 (December 1919): 53; 21 (June–July 1920): 17; 22 (December 1920): 26–27.

33. *AIChE Bulletin* 23 (June 1921): 15–19; 24 (December 1921): 30–35; "Report of Committee on Chemical Engineering Education of the American Institute of Chemical Engineers, 1922," booklet, no pagination; "Survey of Chemical Engineering Education," *Chemical and Metallurgical Engineering* 24 (June 15, 1921): 1047–1053.

34. *AIChE Bulletin* 24 (Dec. 1921): 34.

35. *AIChE Bulletin* 25 (June 1922): 26, 30–32.

36. *AIChE Bulletin* 31 (June 1925): 19–40.

37. For the results of the questionnaire see *AIChE Trans* 19 (1927): 227. For the decision to continue the accreditation program see *AIChE Council Minutes*, March 16, 1927 (AIChE Archival Records, New York).

38. H. C. Parmelee, "Committee on Chemical Engineering Education," *AIChE Trans* 28 (1932): 322.

39. Max Appl, "The Haber-Bosch Process and the Development of Chemical Engineering," in Furter, *Century of Chemical Engineering*, pp. 46, 50; Jean-Claude Guédon, "Conception and Institutional Obstacles to the Emergence of Unit Operations in Europe," in Furter, ed., *History of Chemical Engineering* (Washington, DC: American Chemical Society, 1980), pp. 62–70.

40. Guédon, "Conception and Institutional Obstacles," p. 67, who cites Carl Duisberg, "The Education of Chemists," *Journal of the Society of Chemical Industry*, July 1931, p. 174.

41. Reynolds, *History of the American Institute of Chemical Engineers*, pp. 5–7.

42. C. F. Burgess, "Electrochemistry as an Engineering Course," *SPEE Proc* 10 (1902): 128. For similar complaints along these lines see also Richard K. Meade,

"Why Not 'The American Society of Chemical Engineers'," *The Chemical Engineer* 2 (Oct. 1905): 392, W. D. Richardson, "The Analyst, the Chemist and the Chemical Engineer," *Science* 28 (1908): 401, and "The Chemical Engineer—His Functions and Training," *Chemical and Metallurgical Engineering* 23 (Aug. 25, 1920): 333.

43. H. M., "Industrial Chemistry and Industrial Fellowships," *Chemical Engineering* 13 (1911): 171.

44. B. G. Reuben and M. L. Burstall, *The Chemical Economy* (London: Longman, 1973), p. 19; L. F. Haber, *The Chemical Industry, 1900–1930* (Oxford: Oxford University Press, 1971), pp. 14–17, 26–29, 61, 108–134, 173–184; "Prof. F. Haber on Electrochemistry in the United States," *Electrochemical Industry* 1 (1902–03): 350; and "The German Chemical Industry," *The Chemical Engineer* 8 (1908): 168. Trescott, "Unit Operations," p. 2, notes that unit operations emerged from American mass production industries, especially those involved with metal working and metal making. On p. 10 she notes its link to the 'distinctive' American approach to manufacturing with its mass market mindset of rapid, volume production of inexpensive goods and experience with large-scale design. See also pp. 15, 16.

45. See "The German Chemical Industry," p. 168, on the prominent position of the German research chemist. See also Guédon, "Conceptual and Industrial Obstacles," p. 64, and William J. Hale, "The Immediate Needs of Chemistry in America," *Industrial and Engineering Chemistry* 13 (1921): 463.

46. See Reynolds, "Defining Professional Boundaries," pp. 701–704. See also, for example, "Says Chemistry Should 'Break into Society'" [interview with Ellwood Hendrik], *New York Times Magazine*, 5 March 1916, p. 13. DeLos N. Hicok, a chemist, noted in "Chemists in Public Life" [letter to editor], *Chemical and Metallurgical Engineering* 27 (Oct. 25, 1922): 820, that chemists were paid little, kept at routine tasks, and given little opportunity to apply their knowledge to production and manufacturing.

47. John C. Olson, "Chemical Engineering as a Profession," *AIChE Trans* 28 (1932): 302; see also pp. 308–309 and [Harry McCormack?], "The Young Man in Chemical Engineering," *The Chemical Engineer* 11 (1910): 177.

48. Hugh Griffiths, "London Letter," *Industrial and Engineering Chemistry* 16 (July 1924): 760. The dean of the College of Engineering at Ohio State observed in 1902 that in some quarters there was "actual distrust, as if the chemical engineer was neither chemist nor engineer, but was endeavoring to cover his weakness by a high-sounding title"; see Edward Orton, "The Subdivision of the Field of Chemical Engineering," *SPEE Proc* 10 (1902): 134. Julian C. Smith, *The School of Chemical Engineering at Cornell* (Ithaca, NY: College of Engineering, Cornell University, 1988), p. 1, notes that the chair of the department of chemistry at Cornell maintained that there was no such thing as a chemical engineer.

49. *AIChE Bulletin* 11 (Aug. 1915): 16–17.

50. "The Naval Consulting Board's First Meeting and First Recommendation," *Metallurgical and Chemical Engineering* 13 (1915): 708.

51. For example, the AIChE was not invited to join the Preparedness Canvas in 1916 or the Committee on Commercial Engineering in 1920. See *AIChE Bulletin* 13 (June 1916): 32, and *AIChE Bulletin* 21 (June–July 1920): 18.

52. *AIChE Bulletin*, 24 (Dec. 1921): 53

53. Fred. W. Atkinson, "The Development of the Chemist as an Engineer," *AIChE Trans* 3 (1910): 153. Theodore Jesse Hoover and John Charles L. Fish, *The Engineering Profession* (Stanford, Calif.: Stanford University Press, 1941), p. 163, noted that the early chemical engineering curricula seemed to have been designed by simply selecting courses from the mechanical engineering curriculum and the chemistry curriculum and adding in a number of descriptions of various chemical industries. A popular ditty in early-twentieth-century American colleges was: "Chemical engineering includes most any old thing under the sun that brings in the mon[ey]." See Parmelee, "Chemical Engineering in Industry," *Chemical and Metallurgical Engineering* 29 (Aug. 20, 1923): p. 305.

54. R. T. Haslam, "The School of Chemical Engineering Practice of the Massachusetts Institute of Technology," *Journal of Industrial and Engineering Chemistry*, 13 (1921): 465–468; W K. Lewis, "Evolution of the Unit Operations," *Chemical Engineering Progress* 54, no. 5 (May 1958): 65.

55. "Report of Committee on Chemical Engineering Education of the American Institute of Chemical Engineers 1922," booklet [New York: AIChE, 1922], nonpaginated.

56. In Frank Hemingway, "Organization of Chemical Industries," *AIChE Trans* 10 (1917): 365–366.

57. "The Chemical Engineer of the Future," *Chemical and Metallurgical Engineering* 26 (May 24, 1922): 961, reporting on a conference held that same month.

58. *AIChE Bulletin* 32 (Dec. 1925): 19.

59. As early as June 1926 the Secretary of AIChE reported that from the correspondence he was receiving it was "obvious" that note was being taken of the reports of the Committee on Chemical Engineering Education and that the reports were "having a decided effect on education," and he reported requests for the list of approved schools from men looking for places to send their sons (*AIChE Bulletin* 33 [June 1926]: 18). See also *AIChE Bulletin* 40 (Dec. 1929): 23 ["Many institutions have . . . asked our advice and opinion about courses of study."]; and *AIChE Bulletin* 43 (June 1931): 21, where H. C. Parmelee notes that schools were "deeply interested" in having their names enrolled on the accredited list. See also the discussion of standards in "Accrediting of Institutions Teaching Chemical Engineering by the American Institute of Chemical Engineers," *AIChE Trans* 27 (1931): 402–410, and Parmelee, "Report of the Committee on Chemical Engineering Education, 1925," *AIChE Bulletin* 31 (1925): 19–25.

60. Parmelee, "Committee on Chemical Engineering Education," *AIChE Trans* 28 (1932): 322.

61. H. W. Burr, "Some Features of Engineering Education," *SPEE Proc* 29 (1921): esp. 65–69.

62. Commentary on "Report of Committee to Cooperate with the American Society of Mechanical Engineers," *SPEE Proc* 28 (1920): 130–131. For other expressions of the idea that after 1910 engineering education in America had slipped compared to other areas like agriculture, business, medicine, and law see Wickenden, "Memorandum for the Board of Investigation and Coordination," May 12, 1927 [Wickenden Committee Archival Records] and J. H. Dunlap, "Cooperation Between the Engineering Societies and Educators in Standardizing Engineering Education," *Municipal and County Engineering* 59 (Aug. 1920): 59. H. W. Burr, a prominent consulting engineer, accused the SPEE of supporting the "hopeless jumble" that engineering education had become and criticized it for failing to provide to provide guidance ("Some Features of Engineering Education," *SPEE Proc* 29 [1921]: 65–69).

63. Charles L. Walker, "Origin and History of the Present Study of Engineering Education by the Society for the Promotion of Engineering Education," *Sibley Journal of Engineering* 39 (May 1925): 332–333, 341–342; Charles F. Scott, "Report of the Board of Investigation and Coordination," *SPEE Proc* 33 (1925): 25–27; *Wickenden Report*, vol. 1, pp. 1–12.

64. P. F. Walker to Charles Scott, Dec. 14, 1923; Scott to Board of Investigation and Coordination, Dec. 19, 1923 [Wickenden Committee Archival Records].

65. Wickenden, "Memorandum for Board of Investigation and Coordination," March 4, 1926 [Wickenden Committee Archival Records].

66. "Minutes of the Meeting of the Board of Investigation and Coordination," June 18, 1926; Wickenden, "Memorandum for the Board of Investigation and Coordination," May 12, 1927; Wickenden, "Memorandum for the Board of Investigation and Coordination," Feb. 18, 1928 [Wickenden Committee Archival Records]. Also "Report on the Advisability," pp. 198–213.

67. Wickenden, "What the National Engineering Societies Can Do for Engineering Education," *Mechanical Engineering* 50 (Feb. 1928): 119–124. Wickenden also played around with the idea of the SPEE having a special division of institutional members, with high standards for admission. Membership in this group would then serve as an indirect means of accreditation (Wickenden, "Report on Future Organization and Activity," *SPEE Proc* 36 [1928]: 161–168).

68. "Report on the Advisability" (n. 11 above). See also Wickenden, "Memorandum on the Question of the Advisability and Practicability of Classifying Engineering Colleges," March 13, 1931 [Wickenden Committee Archival Records, Washington, DC].

69. Jackson, *Present Status*, 40. C. F. Hirshfeld, "Engineers' Council for Professional Development," *SPEE Proc* 41 (1933): 358, noted that engineering educators felt it was better to have schools accredited nationally by professional societies than to have multiple lists of accredited lists by state.

70. Parker H. Daggett, "Present Status of Engineering Registration," *SPEE Proc* 39 (1931): 873–874; Jackson, *Present Status*, p. 149.

71. R. I. Rees, "Cooperation of the Society for the Promotion of Engineering Education with Other Engineering Societies," *SPEE Proc* 40 (1932): 36–42; Jackson, *Present Status*, pp. 36–37.

72. Summary of emergence of ECPD can be found in Jackson, *Present Status*, 35–45. See also R. I. Rees, "Cooperation of the Society for the Promotion of Engineering Education with Other Engineering Societies," *SPEE Proc* 40 (1932): 36–42; Hirshfeld, "Engineers' Council for Professional Development," pp. 351–361; Hammond, "Engineers' Council for Professional Development Will Accredit Engineering Colleges," *SPEE Proc* 42 (1934): 390–393.

73. Wickenden, "Memorandum for Board of Investigation and Coordination," March 4, 1926 [Wickenden Committee Archival Records].

74. *AIChE Bulletin* 31 (June 1925), pp. 19–20, 37.

75. Wickenden, "Report of the Director to the Board," Oct. 16, 1925 [Wickenden Committee Archival Records].

76. Wickenden, "Memorandum for the Board, March 4, 1926" [Wickenden Committee Archival Records].

77. Arthur D. Little, "Chemical Engineering—What It Is and Is Not," *AIChE Trans* 17 (1925): 174 [in discussion of the paper]. Note, while this paper was contained in the 1925 volume of the *Transactions*, it contained a report on the Conference on Chemical engineering Education sponsored by the AIChE and held in April 1926.

78. "Minutes of the Meeting of the Board of Investigation and Coordination," April 9–10, 1926 [Wickenden Committee Archival Records].

79. *Wickenden Report*, I, 661.

80. William. P. Kimball, "The Accreditation-Education Interface," *Engineering Education* (May 1968): 1042.

81. Jackson, *Present Status*, p. 154.

82. Hammond, "The Problem of Accrediting Engineering Colleges," *SPEE Proc* 41, 1933, p. 173.

83. Hammond, "The Problem," pp. 174–175.

84. For example, neither Wickenden's "What National Engineering Societies Can Do" (n. 67 above) nor the 1931 SPEE "Report on the Advisability" (n. 11 above) even suggests accreditation by curriculum. The underlying assumption in both pieces seems to be that accreditation would be by school or college of engineering.

85. Parmelee, "Committee on Chemical Engineering Education," *AIChE Trans* 28 (1932): 323.

86. AIChE Council Minutes, August 15, 1981 (AIChE Archival Records, New York). LeBold et al., "Accreditation," p. 182, also noted parallels between the AIChE's program and the ECPD's, and in 1954 Harold L. Hazen, dean of MIT's graduate school, commented that the ECPD's program in its early stages "in some ways paralleled the earlier program of the American Institute of Chemical Engineers"; see Harold L. Hazen, "The Engineers' Council for Pro-

fessional Development Accreditation Program," *Journal of Engineering Education*, 45, no. 2 (Oct. 1954): 101–111.

87. U.S. Dept. of the Interior, *Survey of Land-Grant Colleges, 1930*, p. 844. Undergraduate engineering education at land-grant schools continued after 1930 to be fully representative of American engineering education as a whole. In 1961 Frederick C. Lindvall, dean of Engineering at the California Institute of Technology observed that there was "nothing obvious which distinguishes a Land-Grant engineering school from other State supported institutions" and saw them as being fully integrated into engineering education in the United States ("Evaluation of Resident Instruction in Engineering at Land-Grant Institutions," in *Proceedings of the American Association of Land-Grant Colleges and State Universities*, 75th Annual Convention-Centennial Convocation, v. 2, Nov. 12–16, 1961, pp. 99–102).

The End of "Try-and-Fly"

The Origins and Evolution of American Aeronautical Engineering Education through World War II

Deborah G. Douglas

The fact that some land-grant colleges and universities have departments of aeronautical engineering is coincidence. The Morrill Act of 1862 (and successive pieces of legislation) was a powerful stimulus for the establishment of public higher education, and most consider the law's passage as a key point in the early history of American engineering education as well.[1] However, it is important to note that the principal beneficiaries this support were civil, mechanical and electrical engineering programs.

The goal of the Morrill Act was "to promote the liberal and practical education of the industrial classes in the several pursuits and professions in life. . . ." That objective was both an acknowledgement and a harbinger of the forces of industrialization and urbanization that transformed the United States during the nineteenth century. Not surprisingly then, the engineering education programs that flourished were those related to transportation, communication and manufacturing technologies. Around the start of the twentieth century, the impact of land-grant money on engineering education was greatly diminished because the monetary value had become insignificant in comparison to the patronage of industry and private philanthropists. It would be a mistake, however, to conclude that the federal government was no longer influential.

Federal support for engineering education has been crucial right through to the present day, but no more so than to the development of aeronautical engineering, a discipline that matured over a half-century period between the 1890s and the start of World War II. The story this essay tells is of the first (and arguably most important) engineering discipline to emerge in what could be termed the immediate "post-land-grant" period of federal support for higher education. Instead of federal aid being channeled through state legislatures, it was now granted directly to universities, departments, and even professors. Frequently, financial support came in the form of government contracts.

As important as money was to aeronautical engineering programs, it is critical to note that "aid" should not be defined in narrow financial terms. The federal government was interested in the creation of new knowledge and the formation of a thriving professional community that could support America's military and industrial needs. Subsequent sections of this paper will make clear that the not-so-invisible hand shaping aeronautical engineering education in the United States was the federal government. This federal presence and the fact that it was considered a norm—and, therefore, expected—is one of the most important results of the federal land-grant program of the nineteenth century.

Aeronautical Engineering Education before World War I

Before the First World War, there was no institution of higher learning granting degrees in aeronautical engineering. A handful of schools (and this includes vocational schools) offered a course (maybe two) on various aspects of the technology of flight, but an aspiring engineer, enthralled with the "progress in flying machines," did not have the choice of seeking a baccalaureate or even a graduate degree. Yet there were individuals who identified themselves as "aeronautical engineers." There was a nascent community of scientists and engineers who had turned their sights on the problem of flight even before the Wrights. Two centuries of generalized scientific investigations of fluid motion had cohered into a body of knowledge that had come to be known as aerodynamics in the mid-nineteenth century. Around that time, experimentalists and inventors began building actual vehicles.

Ten months after Orville and Wilbur took turns flying their machine on the sands of the Outer Banks, in August 1904, a thirty-year-old PhD made a short ten minute presentation of a paper titled "Über Flüssigkeitsbewegung bei sehr kleiner Reibung" (On the Fluid Mechanics of Very Small Friction) during one of the "stand-up" sessions at the Third International Mathematical Congress held in Heidelberg. In a span of time only slightly longer in duration than the Wrights' first flights, Ludwig Prandtl introduced the boundary-layer concept. It was a new paradigm for theoretical aerodynamics, every bit as revolutionary as the 1903 Flyer, and it supplied a research agenda and a rationale for building "wind tunnels."[2] Just a few months after delivering his paper, Prandtl was snapped up by Germany's most prestigious research university, Göttingen, where he immediately developed a closed-circuit tunnel (prior to this time, wind tunnels were open tubes or tunnels) and began developing a research center which would produce some of the most significant figures in aerodynamics history—Theodore von Kármán, Max Munk, Adolf Busemann, and Jakob Ackeret.[3]

Prandtl's lab at Göttingen was one of three European research centers that would prove influential in shaping early American aeronautical engineering education. The other two were the National Physical Laboratory in Great Britain and Gustave Eiffel's Laboratoire Aérodynamique in France.[4] The research, reports and textbooks resulting from the work of scientists and engineers at these three facilities shaped the core curriculum for the earliest American courses in aeronautics. Though Americans had invented the airplane, European nations had established a tradition (between a decade and a quarter-century, depending on the country) of supplying government support for aeronautical research.[5] Further, the more extensive and intensely competitive European field of contests, races and exhibitions with rich purses had a salutary and stimulating effect on the technical development of the airplane. Such events had become showcases for national achievement and they prompted a quest for performance—higher, farther, and most importantly, faster—that did not exist to the same degree in the United States.[6]

This fact greatly distressed the small but passionate community of aviation advocates in the United States. To their way of thinking, aviation was the "epic poetry of technological deeds"[7] that could be used for both military and civilian purposes. Within this community, there was a smaller group who believed that the single most important need was for the federal government to establish a national research laboratory. A spirited campaign began around 1910 and after numerous iterations ultimately led to the creation of the National Advisory Committee for Aeronautics (NACA). The process was one that involved close scrutiny of the research facilities in Europe. In 1913 Catholic University professor and aeronautical engineering pioneer Albert Zahm and Navy Lieutenant Jerome C. Hunsaker were sent overseas to survey the laboratories and university programs.[8]

Jerome Hunsaker was a Naval Academy graduate who had been sent to get a master's degree in naval architecture as part of the navy's official postgraduate program in MIT's Department of Naval Architecture. While studying ship propeller design, Hunsaker discovered the works of Gustave Eiffel and in the process became an aviation enthusiast. MIT administrators, especially President Richard Maclaurin, viewed Hunsaker as a gifted engineer and highly esteemed his leadership abilities. In 1913, one year after Hunsaker completed his MS degree and was working at the Boston Navy Yard, Maclaurin requested that the navy detail Hunsaker to MIT for the purpose of developing courses in aerodynamics and aeronautical engineering. To prepare, Maclaurin thought it would be a good idea for Hunsaker to visit Europe. Thus it was that in the summer and fall of 1913, Hunsaker and Zahm made their grand tour.[9]

The elite group of American scientists and engineers which sent Zahm and Hunsaker to Europe had a bias towards European education that ran deeper than the fact that Europeans were more technically advanced than Americans in the field of aviation. Just as American learned and scientific societies had done, educators who established the very first engineering education programs (beginning with West Point) turned to European models. By the mid-nineteenth century hundreds of American students were traveling to Germany to study in the new laboratories and research institutes.[10] The high point in the German university's international reputation came at the turn of the twentieth century, a fact which individuals like Charles D. Walcott, secretary of the Smithsonian Institution and leader of the group promoting American aviation, and others were well aware of.[11]

There would not have been much anyway in the United States for Hunsaker to study. There were a small number of aviation courses, but Hunsaker and his colleagues were not interested in developing a course of study for ingenious mechanics. They wanted a "real" engineering curriculum and to follow the same pattern of development made by mechanical, civil, electrical and mining engineering during the preceding half-century. Engineers interested in the problem of flight would, it was assumed, eventually have their own departments, research facilities and professional societies. There was a path towards professionalization that advocates for aeronautical engineering fully intended to follow. As historian David Noble put it: "Science, like management, gave engineers their identity. The mechanical engineers had emphasized the scientific nature of their discipline to distinguish themselves from mechanics and skilled craftsmen and this emphasis increased over time as scientifically trained engineers made headway against the rule-of-thumb methods of the shop-culture tradition."[12]

If this represented the ambition of individuals like Hunsaker, it is important to keep in mind actual practice. The design and construction of aircraft in the years before World War I involved substantially more experimentation than theoretical investigation. Before 1914, there were about a half-dozen companies building airplanes; their engineers (if the term was used) were generally self-taught. For example, in 1912 Glenn L. Martin Company hired eighteen-year-old Lawrence D. Bell as a mechanic. By 1915, Bell was plant manager.[13] It was not the case that these self-taught engineers were uninterested in scientific discovery or engineering practice. Indeed, the success of the Wrights has been attributed to the fact that they approached the problem of heavier-than-air flight with a systematic method of inquiry that embodied the best engineering practices.[14] While some used the techniques of parameter variation, "try-and-fly" remained the single most important method of experimentation. As long as there was no market for aircraft, the companies which built them remained small, economically-fragile affairs, unable to afford the skills of a university graduate (even if they had been available for hire).

Engineers, as historian of technology Edwin Layton has pointed out, are expensive. An industry had to have attained some measure of economic maturity and be of sufficient scale in order to be able to afford the skills of professional engineers. Thus, the paradoxical situation of aeronautical engineering education in the United States on the eve of World War I was the fact that advocates knew (or at least thought they did) exactly what needed to be done, but the absence of market demand for aircraft squelched these ambitions and forced a slower course of action. This had a number of important implications, chief among them the fact that the federal government would become a critical force in the development of the aeronautical engineering curriculum in the United States.

The MIT example is instructive. Interest in aeronautics at the Institute dates to 1896 when a student modified a ventilation duct to serve as a wind tunnel. In 1910, the Alumni Council recommended that MIT establish a course in aeronautics. The Institute offered its first course in aeronautics in 1913 and the Corporation (MIT's governing body) approved a degree-granting program in aeronautical engineering (to be led by Jerome Hunsaker) starting in the 1914–1915 academic year. In June 1915, MIT awarded its first Master of Science in aeronautical engineering.

But none of this happened in a vacuum. It was not simply a matter of enthusiasm or "knowledge for knowledge's own sake." Aeronautical engineers were "made" because it was thought that there were jobs for them to do. Unlike traditional classical education, no university institutes an engineering curriculum and hires professors without a sense of market demand for the students. In the case of MIT, the market was the United States Navy, which, as noted earlier, had already established a partnership with the Institute to conduct its post-graduate training in naval architecture and construction. It was a lucrative arrangement for MIT, one that the school was eager to expand. Further, President Maclaurin hoped that it would demonstrate MIT's strong desire to be the home of a national laboratory for aeronautics that was being discussed in Washington. Again, it is important to note that the commitment of resources is proportionate to the perceived demand. Until the start of the First World War that demand was quite limited.

Aeronautical Engineering, 1914–1926

World War I changed everything. Though the United States would not enter the war until 1917, the effects of the European conflict on American aviation were swift and profound. Two developments would have special import for the aeronautical engineering profession. One was the military decision to use aircraft in significant numbers as part of the war effort; the other was the establishment of two federal aeronautical research facilities. Market demand would stimulate a nascent industry by creating employment opportunities for aeronautical engineers, while the new laboratories created a vital research agenda and served as the loci of the new technical community. Not surprisingly, the first university programs in aeronautical engineering emerged during this period. Three programs—MIT, Michigan, and New York University—stand out, although in 1926 the number of universities offering courses had grown from 2 to 23 (15 granted either undergraduate or graduate degrees). In this first period, aeronautical engineering manifested a characteristic of all fledgling disciplines—the struggle to define a curriculum; but what is notable is the presence of the federal government.

Federal appropriations for aviation grew in an astonishing fashion, from hundreds of thousands to hundreds of millions of dollars.

Federal Appropriations for Aviation, 1914–1919[15]
(in thousands of $)

	Army	Navy	Post Office	NACA
1914	175	10	—	—
1915	200	10	—	—
1916	800	1,000	—	5
1917	18,082	3,500	—	88
1918	734,750	61,133	100	112
1919	952,305	220,383	100	205

Much of these funds were for the purchase of new aircraft. The tremendous increase in the demand for aircraft had an important effect on aeronautical engineering, although less than one might initially predict. The perceived problem for the aircraft industry was not design but production (hence the entrance of Henry Ford and the automotive industry as consultants). Thus, while the army trained tens of thousands as pilots or mechanics and industry employed equally large numbers of workers, few expressed any concern that there were only two degree-granting aeronautical engineering programs—MIT and the University of Michigan. MIT was a graduate program with a handful of students; Michigan, which granted undergraduate, degrees had 34 students in the fall semester of 1918–1919. Ten students continued into the spring semester, and of those, only two qualified for the bachelor's degree.[16]

While there were many who had degrees in other fields of engineering, in 1914–1918, most who wanted to do aeronautical engineering simply learned on the job. It was widely known that some of the most prominent pioneers in the industry—Wilbur and Orville Wright, Glenn Curtis and Lawrence Sperry—had only high-school educations. But the image and the small numbers of graduates are misleading, as knowledgeable observers of the war credited European technological success to superior research and educational organizations. While virtually all public attention and congressional ire at war's end focused on the production failures of the industry, what concerned the technical community was America's inability to contribute to the technological advancement of the field.[17] This was attributed to the lack of an integrated system of research and development, a problem that many hoped would be rectified by continued expansion of the army's research facility at McCook Field in Dayton, Ohio and the eventual opening of the NACA's laboratory in Hampton, Virginia. (Even before the 1920 dedication, the NACA facility would be named the Langley Memorial Aeronautical Laboratory.)[18]

In fact, the steady growth of McCook and Langley would be an excellent stimulus for aeronautical engineering education programs. Not only did they seek to hire college-educated engineers, but both also issued contracts to universities to conduct research. These two laboratories helped formulate a meaningful research agenda that became the nexus of an emergent technical community.

In 1920, the NACA recruited Max Munk, a gifted student of Ludwig Prandtl. Munk replaced Edward P. Warner (who had resigned as Langley's chief physicist, to return to MIT). Munk had earned two doctorates from Göttingen—one in engineering and one in physics—but postwar anti-Semitism had caused him to consider leaving Germany. Most of his work at the NACA focused on theoretical problems, but Munk was also the creative genius and designer of the Variable Density Tunnel (and later the Propeller Research Tunnel), which began operations in 1922. This wind tunnel transformed the Langley laboratory into a major research facility, on par with the best European facilities.

Suddenly the NACA was in the position of being asked for advice about engineering education. This led the NACA to embrace a new role in encouraging the development of aeronautical engineering programs, in order to support a growing national research enterprise. Its 1921 annual report announced an important new educational outreach initiative. One part of that effort involved developing "problems in aeronautics and answers thereto, which will be distributed among professors in engineering courses at various universities with a view to having the engineering students generally acquire some knowledge of aeronautical engineering."[19]

In 1922, *Aviation* publisher Lester Gardner compiled the first *Who's Who in American Aeronautics.* Of the 896 biographical entries, 132 entries (15%) listed "aeronautical engineer" for occupation. Most of these individuals had earned degrees in engineering (71%) or science (11%). Only 23 did not have a college degree. Out of the total, only 15 had degrees in aeronautical engineering. It was for this reason that the NACA republished an article written by its former chief physicist, Edward Warner, for the *Christian Science Monitor* entitled "Training Aeronautical Engineers."[20] Although the Aeronautical Chamber of Commerce of America listed 16 different colleges and schools offering courses in aeronautics,[21] Warner noted: "The field is so new that its teaching has hardly become systematized as yet, but some measure of agreement is at least being reached on the essential elements that an aeronautical course should include, and on the requisites of preliminary preparation."[22]

The demand for aeronautical engineering graduates was severely limited, which led Warner to conclude that aeronautical engineering was best restricted to graduate instruction: "the production of a few highly qualified men rather than to the less thorough training of a much larger number."[23] Warner stated that students interested in research should study experimental physics or other research fields as undergraduates, while those interested in aircraft design should take civil or mechanical engineering or naval architecture. (He was biased towards civil engineering unless a student was interested in propulsion.) He thought all students needed a "bent for mathematics," as "the study of the theory of design, and especially of the theories of wing action and of air flow around bodies which have recently been developed, requires mathematical attainments beyond those exacted of the ordinary practitioner in any other sort of engineering work."[24]

Warner had two further recommendations. Aeronautical engineering students should learn to pilot an aircraft and they should spend time working in an aircraft factory. "It is unnecessary to emphasize," Warner writes, "that no opportunity should be lost to acquire practical experience as well as that knowledge which can be gained from lectures and books."[25] Warner's recommendations that students needed practical as well as theoretical training represented an ideal—although it was an ideal that many thought was the best possible path. Actual curricula in most instances focused on theoretical training. For example, both the California Institute of Technology and Stanford University were home to some of the most prominent aeronautical engineers or researchers (Harry Bateman and Albert Merrill at the California Institute of Technology; William Durand and Everett Lesley at Stanford) and both offered graduate courses in aeronautics, but neither had a formal course of instruction or department in this 1914–1926 period.

Not everyone agreed with Warner's idea that aeronautical engineering should be restricted to graduate students. In 1916, the University of Michigan listed three courses in aeronautical engineering, the start of a four-year program of study.[26] Initially the technical content of the curriculum was influenced by the French aeronautical engineering curriculum taught at the University of Paris by Lucien Marchis. In response to a letter from the editor of *Aerial Age Weekly,* dean of engineering Mortimer Cooley summed up the university's general objective as follows: "Our aim is to teach the theory of aeroplanes and to enable students to secure positions in manufacturing plants."[27]

By war's end, the major influence had shifted to the needs of the military. In the spring of 1917, Michigan professor Felix Pawlowski spent a semester's leave working as an aeronautical engineer for the Army Air Service. He returned in the fall to establish a special course at the university that would help students prepare to become army pilots. After the war, Michigan sought to expand its aeronautical engineering program with a special emphasis on research. By 1922 plans for a new engineering building included a wind tunnel and lab space.[28]

Michigan inspired the formation of other undergraduate programs, most importantly one at New York University. NYU offered its first course in elementary aerodynamics in the spring of 1923. The popularity of the course led its instructors, Collins Bliss and Alexander Klemins, to propose a four-course program for seniors majoring in mechanical engineering. The program received approval in 1924, contingent on raising $500,000 in outside funds to support it.

The university formed a special committee in March 1925 that included Harry F. Guggenheim, the aviation-minded son of Daniel and Florence Guggenheim. Harry did not think much of the idea of a large public campaign. At the meeting he simply argued in favor of beginning a search by approaching private philanthropists who could fund the entire project. What eventually became clear was that Harry thought this was a project his father ought to support.

Harry was an aviation enthusiast, learning to fly in 1917 for fun. He later served in the navy as a pilot in World War I and became friends with many men who later would rise to be leaders in the aircraft industry and federal government. Through Harry, Daniel Guggenheim was introduced to aviation. To Daniel Guggenheim's way of thinking, aviation's greatest potential was not as a military weapon but as an instrument of peace. With this powerful notion in mind, Daniel Guggenhiem decided to fund the NYU program.[29]

The $500,000 gift to fund the Daniel Guggenheim School of Aeronautics was announced in June 1925, and groundbreaking for the new laboratory and classroom building occurred the following October. Daniel Guggenheim announced: "Were I a young man seeking a career in either science or government, I should unhesitatingly turn to aviation. I consider it the greatest road to opportunity which lies before the science and commerce of the civilized countries on the earth today."[30] It was a catalytic moment, as just a few months later it led to the creation of the Daniel Guggenheim Fund for the Promotion of Aeronautics—a $2.5 million fund for the support of aviation, especially for the promotion of aeronautical education.[31]

Aeronautical Engineering Education, 1926–1933

The Guggenheim Fund was ostensibly a private foundation, but the intent of Daniel and Harry Guggenheim was that its work complement the efforts being made by the federal government. In fact, Daniel Guggenheim first proposed the idea during a lunch meeting with President Calvin Coolidge and Secretary of Commerce Herbert Hoover. The official announcement of the Fund's creation was made in a letter to Hoover in January 1926. Many of the Fund's trustees were drawn from the ranks of government. The 1926 *Annual Report of the National Advisory Committee for Aeronautics* noted the extremely close cooperation between the Committee and the management of the Fund.[32] That relationship was intended to be a model of Hoover's vision of the "associative state," a cooperative venture between public agencies and private organizations to serve the public good.

One of the priorities that all shared was a desire to promote engineering education and research in American universities. The federal government could award research contracts, shape a research agenda, and even nurture a technical community, but it could neither supply funds for new faculty nor construct the necessary infrastructure for college campuses to become meaningful research centers. What the Guggenheim family had done for New York University, the Guggenheim Fund would now do for six more schools. The California Institute of Technology, the University of Michigan, the Massachusetts Institute of Technology; the University of Washington, Stanford University, and the Georgia Institute of Technology all received grants (of varying sizes) from the Fund to build wind tunnels, classroom and laboratory space as well as funds to support hiring new faculty. In the case of the latter, the funds were dispersed annually for a ten-year period so as to permit the university the time needed to endow such positions permanently.[33]

The Guggenheim Fund generated enormous excitement within the aviation community. Daniel Guggenheim's gift of $2.5 million (plus the $500,000 he had already given NYU) was a very large sum, equaling about 6.4 percent of the $39 million appropriated by the U.S. government for aviation in 1925-1926.[34] Thus, the Fund's establishment in conjunction with the passage of two major pieces of federal aviation legislation—the Kelly Air Mail Act of 1925 and the Air Commerce Act of 1926—and a series of record-setting flights culminating with Charles Lindbergh's solo crossing of the Atlantic in May 1927 marked a saltation in American aviation in general, and aeronautical engineering education in particular.

These events coincided with a suddenly booming U.S. economy. Aviation was seen as symbolic of progress and modernity, the perfect investment for the future. From 1927 to 1932, historians estimate that the public pushed the value of aviation securities to between $850 million and $1 billion dollars on the New York Stock Exchange. The price/earnings ratio was 100 to one based on estimates of net total earnings of aviation firms between $8 to $10 million.[35] The impact of all this cash was immediate and profound. Flush with capital, imbued with almost universal public enthusiasm and interest, and governed by meaningful regulation, the aviation industry immediately responded with expanded production.

U.S. Aircraft Production, 1926–1933[36]

	Total	Military	Civil
1926	1,186	532	654
1927	1,995	621	1,374
1928	4,346	1,219	3,127
1929	6,193	677	5,516
1930	3,437	747	2,690
1931	2,800	812	1,988
1932	1,396	593	803
1933	1,324	466	858

Not surprisingly, there was an efflorescence of manufacturing companies in 1928 and 1929. For this study of aeronautical engineering education, the more significant trend is the wave of consolidations and the formation of several large holding companies. The perception was that these new companies could afford and would need to hire college-educated engineers.

The Aeronautical Chamber of Commerce of America was the leading promoter of such ideas. Announcing the half-million dollar gift to New York University and the creation of the Guggenheim Fund in its 1926 year book (reporting on events in 1925), the Chamber did a quick survey to determine the scale and scope of aeronautical engineering education in the country. Beyond MIT and Michigan, only Stanford and the University of Washington offered courses in aviation.[37]

The following year, the results from a much more comprehensive survey sent to 500 schools were published. Twenty-three schools gave some kind of instruction in aeronautics, although only 5 offered a course leading to a degree in aeronautical engineering. Fourteen schools claimed to have an aeronautical laboratory or equipment to conduct aeronautical research although only 8 schools said they were actually doing any research work. There were 26 graduate students taking aeronautical engineering, 96 undergraduates were enrolled in aeronautical engineering majors, and 163 undergraduates who were in "regular" (mainly mechanical or civil) engineering courses elected aeronautical subjects.[38] The Chamber was eager to see the numbers increase as it viewed these statistics as important indicators of the general health of the aviation industry. Brief accounts of various university programs, including information about the types of research facilities and equipment, became a standard feature of the year book but they were little more than uncritical (often inaccurate) descriptions.

By contrast, the surveys of aeronautical engineering education conducted by the Association of Land-Grant Colleges and Universities were much more sober. Not surprisingly, aeronautical engineering was not included on the list of "subjects of instruction allowed" under the Second Morrill Act (1890), according to the *Digest of Rulings of the Secretary of the Interior,* December 7, 1900 and May 23, 1916.[39] By 1929, however, several land-grant schools were offering courses in aeronautics with 569 students enrolled. The question arose as to whether or not the Association should promote this. It was not that there was any doubt that aeronautical engineering, as had been the case with other new engineering disciplines such as electrical engineering, would be considered a legitimate course under the "mechanic arts" heading. Rather, the concern was whether or not there was a sufficient demand for aeronautical engineers to justify encouraging other land-grant schools to add it to their curricula.[40] "Fundamentally," wrote Homer Dana in his 1930 report to the Committee on Aeronautics, "a state educational institution is, and should be, more or less conservative in its expansion policies—better to advance slowly and surely than to be forever retracing one's steps from a false position."[41]

Right from the start, it was clear to members of the Committee that further encouragement of aeronautical engineering programs was a "false position." Robert Spencer, dean of the University of Delaware, explained why he had decided against establishing an aeronautical engineering program at Delaware:

> . . . aeronautical engineering, as we all know, has a great romantic appeal to our students and to the friends of the University, many of whom have not made a careful study of the subject. It is one duty of a dean to be progressive, to be a booster. It is another duty of a dean to make a careful study of subjects as they

come up and to consider them soberly. Sometimes his judgement will impel him, much against his wishes, to say, "Not yet."[42]

Richard H. Smith of MIT added that it was important to recognize that aviation was not a normal industry. The tremendous growth in commercial aviation was a consequence of federal investment. "These lines are not what we call subsidized, we dislike that term, but they are supported by the equivalent of a subsidy, and for that reason have been able to go into operation on what might appear to most of you as a profit-making basis." Smith was emphatic about the point that there were too many engineers on the market. While some might blame the Depression, Smith felt the main cause was overproduction.

> What we need in the industry of aeronautics is fewer and better schools and fewer and better engineers. Those two things should be the slogan all over the country. I seriously doubt the wisdom of any school during the next three years adding a department of aeronautical engineering. I say that with no note of pessimism, but simply from a conservative interest in the advancement of aeronautics, and in the health of the profession.[43]

This mindset—fewer and better schools and fewer and better engineers— was strongest in the schools receiving Guggenheim grants and weakest among those universities that needed a strategy to keep up enrollments during the Depression. Thus, Stanford created graduate programs and enrolled very small numbers (5 or 6 per year) of highly qualified students, whereas Minnesota enrolled 229 students the first year of its degree program (1929), 237 in 1930–31; and 225 in 1931–32. The Minnesota course catalog stated that "The aeronautical engineering course is similar to mechanical engineering. The fundamental studies are the same. As a result, the graduates in aeronautical engineering should be prepared to enter various branches of the mechanical engineering field if, for any reason, they should prefer to do so."[44] It would appear to have been a good strategy, as the 1932 alumni directory for the College of Engineering and Architecture lists 44 aeronautical engineering graduates; of the 24 individuals listing their place of employment, only one was connected with aviation.[45]

Minnesota established its aeronautical engineering program in this aviation boom period. Aerodynamics was taught by Charles Boehnlein of the mechanical engineering department, and the popular response to these classes (offered during both the day and evening) in 1928 led Ora M. Leland, dean of the College of Engineering, to invite John D. Akerman, the chief engineer at Mohawk Aircraft Company in Minneapolis, to give a course in airplane design. The University established a department of aeronautical engineering in the fall of 1929, naming Akerman as its head.[46]

Akerman was born in Latvia and studied aeronautics at the Royal Technical Institute in Moscow. Until the Russian Revolution, Akerman had been a pilot in the Imperial Russian Air Force. In 1918 he came to the United States and entered the University of Michigan. He graduated from Michigan's aeronautical engineering program in 1925. Minnesota, like Michigan, focused on its undergraduate program and there were similarities in curriculum and willingness to cater to student interests. The Depression stimulated a vigorous discussion of whether or not the University was producing too many engineering graduates. Dean Leland vigorously denied this and claimed there would always be jobs for graduates. He opposed arbitrary restrictions, stating that as long as there were adequate facilities, equipment, and staff, the university should admit as many students as could be accommodated without sacrificing quality of instruction.[47]

The most significant influence on the Minnesota curriculum was the U.S. Navy. In fact, its sophomore year course served as the ground school for the Navy's Air Reserve Corps. Many Minnesota graduates went on to take flight training at the Naval stations at Great Lakes and Hampton Roads and were commissioned in the Naval Reserve. This relationship antedated the establishment of the department and helped sustain enrollment.[48]

The Minnesota emphasis towards preparing students to become pilots rather than engineers was antithetical to the Guggenheim Foundation Board. Flight training was an important objective, but not one that needed to be part of a university's aeronautical engineering degree program. In contrast to Minnesota was Stanford University. Stanford began aeronautical engineering research in 1916 with the pioneering propeller research studies of William Durand and Everett P. Lesley. For a decade, the National Advisory Committee for Aeronautics supported this research. Lesley taught an aeronautics laboratory course each year. and beginning in 1922 Durand began teaching a theory of flight course for advanced students in the mechanical engineering department. Stanford required five years of study to earn a mechanical engineering degree. Thus, it was not surprising that when the university submitted a grant proposal to the Guggenheim Foundation, it was for a six-year program—two full years of graduate study plus a thesis.[49]

The curriculum Durand designed for this new program was deeply influenced by the developments at the NACA and the military's engineering research laboratories, as well as knowledge of the various European facilities. Durand did not want to create a center for aerodynamics research, however. Stanford's program was more broadly conceived as training aeronautical engineers (meaning the emphasis was on design), thus Durand's (and his colleague Lesley's) desire was to hire topnotch researchers in the fields of aerodynamics and aircraft structures. Lesley would continue to teach the laboratory research courses, but since Durand was now officially retired (he would continue to be active professionally for another two decades), the university would need to hire a replacement to teach aerodynamics and other related courses.[50]

Durand and Lesley knew exactly whom they wanted: Alfred S. Niles and Elliott G. Reid. Niles, along with his colleague Joseph Newell, had transformed McCook Field into the locus of intellectual creativity in aircraft structures research. Shrinking budgets for military aviation made the Stanford offer appealing to Niles (Newell would be hired by MIT), who had just left McCook to work as a stress analyst for Consolidated Aircraft Corporation. Elliott Reid worked at the NACA Langley Memorial Aeronautical Laboratory. While the NACA's budget was expanding, the talented aerodynamicist Reid felt constrained and unappreciated at Langley. Durand, who served on the NACA's main committee, had noticed Reid and shared Lesley's assessment that he was a researcher of exceptional promise.[51]

Niles and Reid began teaching at Stanford in the fall of 1927. In March 1930 an article in the *Stanford Illustrated Review* noted that thirteen of sixty-six graduate students in the School of Engineering were enrolled in the aeronautics program, and six had already begun thesis research. Stanford's close ties with government labs were as strong as ever, with reports of three research projects funded by the NACA published in 1930.[52] The entire Stanford faculty infused their classroom lectures and problem sets with material derived from the ongoing research efforts at government labs. For example, in 1929 Niles and his former McCook colleague Joseph Newell published *Aircraft Structures,* the first aircraft structures engineering textbook.[53] Niles and Newell acknowledge a heavy debt to the army Air Corps publication *Airplane Design,* as considerable material and many illustrations were drawn from that earlier work.[54]

Stanford was very proud of its program. The Guggenheim grant had supplied an excellent research facility and talented faculty. Graduates were being hired for responsible positions in industry and government (in spite of the terrible economic conditions). The NACA continued to give grants and contracts to the university, which Stanford took as endorsements of the quality of its research. The overall picture of aeronautical engineering education was not quite so rosy. The Aeronautics Committee of the Association of Land-Grant Colleges and Universities 1931 survey of airplane and airplane-engine factories found that none had employed any graduates from either the 1930 or 1931 classes. Further, while these factories had come to value a college degree, especially in aeronautical engineering, the survey found that there was an "almost universal complaint that the young engineering graduate does not have enough practical knowledge of shop methods, machinery, and structural parts. . . ."[55] They wanted engineers that knew all the parts of the plane, could rig it properly, make repairs, and improvise solutions to technical problems.

There was an emerging controversy in the aeronautical engineering community between engineers identified as "theorists" and those deemed "practical men." On the one hand, this development could hardly have been unexpected. As collegiate programs in engineering developed in the nineteenth century, bitter struggles over training and credentials had been the norm. Attending the Pacific Coast Aeronautics meeting of the American Society of Mechanical Engineers, Professor F. W. Candee reported that "The industry's main objection to research carried on in the schools is that it requires too long to get results. Test pilots, using cut-and-try methods are therefore often resorted to, in order to obtain results quickly."[56]

Industry was not fully satisfied with the graduates of aeronautical engineering programs. Advocates for either side could point to individuals that had made their mark making it difficult to argue decisively in favor of university training. For example, in 1925 Douglas Aircraft hired two men, James H. "Dutch" Kindleberger (self-taught) and Arthur Raymond (Harvard, MIT), who swiftly rose to become the company's top engineers. By 1933, however, this had changed. The emergence of a new class of commercial air transports—the Boeing 247, Lockheed L-10 *Electra,* and Douglas DC-1 and DC-2—heralded enormous changes in the aircraft design process and a maturation of the aeronautical engineering community. While there are plenty of anecdotes associated with these vehicles that suggest last-minute fixes and design decisions made primarily on the basis of intuition (as opposed to analysis), these were "engineered" airplanes. Yet it was not well understood that the minds that conceptualized them followed the habits of thought and principles taught in aeronautical engineering programs, rather than by "try-and-fly" methods of experiment and design. As Georgia Tech professor Montgomery Knight wrote in a script for a radio talk in 1932, "the activities of the aeronautical engineer are somewhat of a mystery to the average person."[57] That "mystery" would be dispelled as engineers from government, industry and academia argued about the proper training of aeronautical engineers and other technical workers in aviation for the rest of the decade.

Aeronautical Engineering Education, 1933–1940

Amidst the negative news of the Depression, aviation proved to be a source of sustained popular interest. Record-setting flights and races attracted enormous attention and Americans lavished affection on their pilot-heroes. So when Edward Elliott, president of Purdue University, chanced to meet Amelia Earhart at the *New York Herald Tribune*'s Conference on Current Problems in September 1934, he immediately seized on the idea of inviting Earhart to Purdue. Elliott was a staunch advocate of women's education and he thought Earhart would be an outstanding role model for Purdue's growing number of female students. At the same time, aeronautics was a new interest for the university and Elliott hoped that Earhart's presence would draw the aviation industry's attention (and grants) to the school. Though he was fiercely opposed by most of the engineering faculty at Purdue, Elliott, with the support of many alumni, persisted.[58]

The Purdue faculty opposed the Earhart appointment because of a profound prejudice against women entering the engineering profession. Aeronautical engineering faculty at schools like Caltech, MIT, Michigan and Georgia Tech were opposed because they felt it sent the wrong message to the public about their profession. Montgomery Knight aptly summarized the frustration he and his colleagues felt in an article written for Georgia Tech students:

". . . but don't you teach flying?" The gentleman who was asking this question was sitting in my office the other day. He had come in to see me about having his son enter Georgia Tech as a prospective student in Aeronautical Engineering. "No," I replied, "we do not teach flying for several reasons, chief of which is the fact that flying is an art, while aeronautical engineering is a science."[59]

Engineers were neither pilots nor technicians. Knight, a NACA research engineer at the Langley Memorial Aeronautical Laboratory, had been recruited to head the new Guggenheim School of Aeronautics, established in 1930. His vision was curriculum that prepared students for one of two career tracks. The airplane design option would prepare students to work for aircraft manufacturers. The aeronautical research option would train students to work in government or industrial research laboratories. As stated in the Georgia Tech catalog, the purpose of the new school was "the training of aeronautical engineers, the prosecution of aeronautical research, and co-operation with other agencies devoted to the cause of furthering air transportation with particular reference to the South."[60]

Aeronautical engineering research was not the same thing as flight research, however. Knight was constantly writing articles, giving radio talks, and making speeches to explain that the real work of aeronautical engineering was in a lab and that the the single most important research tool was the wind tunnel. The testing of small models (to scale) of aircraft in a wind tunnel was much less costly, infinitely safer, and perhaps, most importantly, a much faster way to determine the success (or limitations) of a particular design idea.[61] Knight had attended MIT as an undergraduate (his BS degree was in electrical engineering) and worked at Langley during one of its most exciting periods of wind tunnel construction. This experience profoundly shaped his educational philosophy, but Knight was equally drawn to the extraordinary developments in this field at the California Institute of Technology.

One of the first letters Knight wrote upon arriving in Atlanta was to Clark Millikan at Caltech. Should Georgia Tech develop an undergraduate degree program? What advice could Millikan provide regarding a graduate program? How soon could graduate students do useful research?[62] In Millikan's absence, Professor Arthur L. Klein responded stating that at Caltech they did not believe undergraduate education was worthwhile. Further, he noted that during a recent NACA meeting on educational research, Professor Pawlowski of Michigan and Professor Klemins of NYU had stated that they had been unable to place all their graduates. As for a graduate curriculum, Klein advised that the focus be on aerodynamics and structures. Under supervision a second-year graduate student could do research that was of a quality sufficient for industry needs.[63] Knight responded appreciatively and indicated that Klein's thoughts were similar to his own, but he would quickly find he lacked the institutional support to replicate the Caltech model.

The aeronautics program at Caltech was both extraordinary and anomalous. Robert Millikan's vision was to create a research center in aeronautics at Caltech. The Nobel Prize winner in physics had tremendous ambitions for the modest technical college that hired him in 1921. Caltech had employed Harry Bateman since 1917 and Albert A. Merrill since 1918, so there was some aeronautics work underway when Millikan began to observe the growing aviation industry in the Los Angeles region. What Millikan wanted was to create a research center that would rival the best European labs. For him, Ludwig Prandtl's facility at Göttingen represented the pinnacle of theoretical and experimental work in aerodynamics and it was precisely the German model he hoped to bring to the United States. When Millikan began soliciting the Guggenheim Foundation for money, this was the vision he put forth.[64]

Initially, Harry Guggenheim was not persuaded by Millikan's proposal, but his 1926 visit to various European labs changed his mind. Millikan's own stature as a scientist, as well as the reputations of the researchers he had recruited to come to California, convinced Guggenheim that Caltech was the most promising site for an "American Göttingen." The key would be finding "a scientist of ability bordering on genius."[65] Millikan supplied a list of possible candidates to the Guggenheims, but only Prandtl and his star protégé Theodore von Kármán were considered viable choices. Millikan wanted the younger von Kármán; the Guggenheims favored Prandtl. After meeting von Kármán, however, the Guggenheims were deeply impressed.[66]

It would take three years to persuade von Kármán to come to America (he did make two extended visits to Caltech in the interim), but finally in the fall of 1930 his tenure began. As predicted, it altered the American aeronautical engineering landscape and vaulted Caltech into the position of the preeminent educational institution in the field. That reputation continued to grow rapidly during the 1930s as Caltech became a center for theoretical study and contract investigations for the aircraft industry. Caltech's close association with Douglas Aircraft Company and the extensive testing of the DC-1 design concepts in the wind tunnel and labs proved especially significant.[67]

When it came to the education of students, Caltech offered only a postgraduate program. Applicants were expected to have completed a four-year program in mathematics, physics and engineering before starting the two-year master's course.[68] The curriculum was organized along three lines.

1. A comprehensive series of theoretical courses in aerodynamics and elasticity with the underlying mathematics and mechanics.
2. A group of practical courses in airplane design.
3. Experimental and theoretical researches on
    (a) the basic problems of flow in real fluids with regard to the scientific foundations of technical hydro- and aero-dynamics;

(b) practical problems in aerodynamics and structures, especially as applied to aeronautics.[69]

Until the late 1930s, Caltech did not admit many students. Then, as the demand for aircraft began to pick up (coupled with increasing belligerence in Europe), there was a similar increase in demand for aeronautical engineers. Clark Millikan responded negatively to a desperate plea from Montgomery Knight for a suitable Caltech PhD to fill a vacancy at Georgia Tech, as "The current boom in the aircraft industry has snapped up all of the available aeronautically-trained personnel of whom I have any knowledge. . . ."[70] In response, Caltech dramatically increased the number of students it admitted to its regular programs and added special one-year courses for army and navy officers assigned to graduate school for advance study in aeronautical engineering.[71]

U.S. Aircraft Production, 1933–1940[72]

	Total	Military	Civil
1933	1,324	466	858
1934	1,615	437	1,178
1935	1,710	459	1,251
1936	3,010	1,141	1,869
1937	3,773	949	2,824
1938	3,623	1,800	1,823
1939	5,856	2,195	3,661
1940	12,813	6,028	6,785

It is important to note that Caltech was not typical either in its educational environment or its relations with industry. As an educational institution it was unique; as a research laboratory its only rival was the NACA's Langley Laboratory, although the comparison is difficult to make given the vastly different resources and research objectives. Most university aeronautical engineering programs had more complicated relations with industry. J. L. Atwood, vice president of North American Aviation, noted that "in industry, excluding teachers in universities and governmental agencies, very few aerodynamics specialists are employed. . . . These men are necessary, but should be exceptional both in qualifications and training."[73] Atwood went on to observe: "My feeling is that aeronautical engineering courses are too 'fancy,' trying to incorporate too much of the abstract science of aerodynamics. . . ." For Atwood the essentials were an ability to "on any occasion derive any of the basic formulas of analytical geometry, calculus, basic stress formulas, basic deflection formulas, and retain this knowledge at all times."[74]

Atwood offered his opinions as part of a survey of aeronautical education. The Institute of Aeronautical Sciences, the newly founded professional association for aeronautical engineers, asked Georgia Tech's Montgomery Knight to give a paper on this topic at its 1939 annual meeting. Knight sent letters to twenty-one leading manufacturers and airlines. Sixteen responded, most with extremely detailed and candid answers to questions about the educational qualifications of various categories of employees (workers in engineering departments, shops, sales, service, dispatching, traffic, and pilots); the success (or failure) of aeronautical educational institutions; the best venue for vocational training; and ways to improve industry-academia relations. Analysis of the responses proved useful to Knight and his contemporaries, but they also indicated that for the first time there was a broad consensus within the aviation community about the role and training of aeronautical engineers.[75]

On the one hand, Knight found that the curricula of the nation's various aeronautical engineering and vocational schools did cover the industry's stated requirements. "Moreover," he wrote, "there is a striking uniformity in the corresponding curricula which leads one to believe that the schools of aeronautics have passed through the evolutionary period." The responses to the survey did reveal that industry believed aeronautical engineering programs needed to give students "a more practical attitude toward the solution of engineering problems."[76]

The term "practical" no longer meant the experimentations of an untrained tinkerer or a mechanic who had some calculus and drafting, however. The objective of a degree program in aeronautical engineering was "to train men for engineering groups of airplane manufacturing plants or engineering maintenance sections of large operating companies," wrote Professor Shatswell Ober of MIT. Ober made a clear distinction between engineers and technicians: "the engineer ultimately plans and guides while the technician executes (often brilliantly)."[77] Industry representatives did not feel that at the entry-level it was useful to draw the lines so sharply. Of the four hundred people in the Douglas Engineering Department, they only needed a half-dozen aerodynamicists of the type trained at Caltech or Stanford and three or four dozen trained at the other Guggenheim schools to do the most difficult structural analysis. The rest were draftsmen. Douglas and all the other respondents felt the universities were not doing enough to prepare students so that they would be immediately useful after graduation.[78]

There was some skepticism that the representatives of industry really meant what they were saying. Shatswell Ober argued that "weeks and weeks spent on the details of particular rules, say of the Civil Air Authority for stress analysis, may give a man superficial knowledge of procedures of immediate value to an employer, but, unreinforced will leave him helpless if the answer for a problem is not found in the rules."[79] His survey of a dozen aeronautical engineering programs revealed that the distribution of courses was as follows:[80]

Foundational Courses:

- Preparation for leadership courses (non-technical courses in the humanities and social sciences) ranged from 5–22%, with an average of 15%.
- General science and mathematics courses (spec. physics, chemistry and calculus) ranged from 18–33%, with an average of 28%.

General Engineering Courses:

- Schools varied widely in the departments from which these courses were drawn, but all included courses in drawing, applied mechanics, thermodynamics and shopwork.

Aeronautical Engineering Courses:

- Professional aeronautical courses ranged from 23–32%, averaging 26%. Of these, aerodynamics accounted for an average of 8%.

The surveys conducted by Ober and Knight suggest that the immediate needs of business were just that: short-term needs based on a sudden increase in the market for their goods. Industry personnel officers and the heads of engineering departments might wish for graduates with better drafting skills, but virtually all wanted to hire students trained in engineering. The educational director for Transcontinental & Western Air, Inc., J. W. Vale, Jr., put it succinctly in his reply to Knight. If an individual wanted to work in TWA's engineering department, he "must be a graduate from an accredited engineering school."[81] There was no mention of needing a pilot's license and no requirement that the engineer be able to repair an engine. The era of "try-and-fly" was over.

Conclusion

The "theory" and "practice" tension perceived by aeronautical engineers during the late 1930s was an indicator of the changing historical context of the airplane. In the early 1900s, the context in which Americans understood the significance of the Wright Brothers' invention was largely one of poetry and dreams. The fact that the airplane was invented at this time as a consequence of the same powerful forces of industrialization that were remaking the nation was not popularly understood. Four decades later this had changed. On the eve of World War II, the context of the airplane was a specific set of applications for commercial and military purposes that required a sophisticated and costly system of infrastructure. One cannot divorce an understanding of engineering education from this fact. In other words, what was adequate training in one context was wholly inadequate in another.

As a large technological system is created, there is a corresponding need for credentialed experts to make judgments that become the basis for investment. How fast? How frequent? How safe? These were questions that shaped the American air transportation system and they were the problems that engineers were asked to solve. It was not simply about a single machine or saving a life; it was about a system. "Try-and-fly" ended not because there was no value in practical experimentation or that the stakes were too high for an individual to bear, but because the demand for increasingly precise answers to the questions above required individuals—operating within an industry, university, or government environment—

who could give rigorous answers with the imprimatur of "science" to give legitimacy and authority. The scope of the work, as much as the scale, had been inextricably altered.

It was the federal government that gave greatest shape to this system and thus the role and education of aeronautical engineers. For example, as the federal government began imposing increasingly stringent airworthiness standards in the 1930s (regulations which governed the minimum safety and design standards which all aircraft had to conform to in order to be allowed to fly), the industry was compelled to adopt design practices that necessitated employing individuals with a particular set of skills. Those standards emerged from the research investigations of engineers at various government laboratories and agencies, so it is not surprising that the perception was firmly established that aircraft, engines, airports, and the other technologies of aviation should be designed by engineers.

That the federal government could determine expertise in science and technology was long accepted. By the start of the twentieth century further refinement of the definition of a science and technology expert meant someone with academic training. The U.S. government was the first and most important customer for the aviation industry. Predictably then, aeronautical engineering programs began to take hold in various colleges and universities. Over time the process of professionalization evolved, and by the start of World War II an educational hierarchy with specialized training for engineers, technicians, and pilots had emerged. The federal government played a major role in defining the education for each of these areas of specialization. Today we speak of "engineering" or "designing" new airplanes, not "inventing" them. The people that do this work may or may not have been trained at a land-grant institution, but their educational experience, the fact that a degree is required, and the shape of their profession are important and enduring legacies of the land-grant legislation.

Notes

Assistance from the following archives and libraries is gratefully acknowledged: University Archives, University of Illinois at Urbana-Champaign; Bentley Historical Library, University of Michigan; Archives and Special Collections, Massachusetts Institute of Technology; University of Minnesota Archives; Stanford University Archives; California Institute of Technology Archives; Boeing Corporation Historical Archives; NASA Langley Research Center History Archives; Archives, Library and Information Center, Georgia Institute of Technology; National Archives and Records Administration, Mid-Atlantic Region; National Association of State Universities and Land Grant Colleges. This essay was prepared as part of a contract history project for the National Aeronautics and Space Administration, Langley Research Center.

1. Act of July 2, 1862 (Morrill Act), Public Law 37-108, 12 STAT 503. Other major acts include: Act of August 30, 1890; Hatch Act of 1887; and Smith-Lever Act of 1914. There have been various amendments as well as separate rulings and opinions associated with these pieces of legislation.

2. The term "wind tunnel" was introduced by Gustave Eiffel in 1910, although it did not come into English use until 1913, when Jerome Hunsaker translated Prandtl's paper for the NACA. John D. Anderson, *A History of Aerodynamics and Its Impact on Flying Machines* (Cambridge: Cambridge University Press, 1997), p. 275.

3. Anderson, pp. 258–260.

4. A. Klemin and T. H. Huff, "Course in Aerodynamics and Aeroplane Design, Part I., Section (I., 'Modern Aeronautical Laboratories,'" *Aviation* 1 (1 August, 1916): 9–16. Other important centers for aeronautical research included the Royal Aircraft Factory in Great Britain; the Institut Aérotechnique de l'Université de Paris in France; and the Deutsche Versuchsanstalt für Luftfahrt zu Adlershoff in Germany.

5. Tom Crouch, "Blaming Wilbur and Orville: The Wright Patent Suits and the Growth of American Aeronautics," in Peter Galison and Alex Roland, eds. *Atmospheric Flight in the Twentieth Century* (Dordrecht: Kluwer Academic Publishers, 2000), pp. 293–295.

6. Ibid.

7. Robert Wohl, *A Passion for Wings: Aviation and the Western Imagination, 1908–1918* (New Haven: Yale University Press, 1994), p. 29.

8. Alex Roland, *Model Research* (Washington: NASA, 1985); see chapter 1, "The Quest, 1910–1915," pp. 1–25.

9. Shatswell Ober, "The Story of Aeronautics at M.I.T., 1895–1960," April 28, 1965, p. 6, Mss Collection AC 43, Box 17.27, Institute Archives, Massachusetts Institute of Technology, Cambridge.

10. Alexandra Oleson and John Voss, "Introduction," *The Organization of Knowledge in Modern America, 1860–1920,* edited by Oleson and Voss (Baltimore: Johns Hopkins University Press, 1979), pp. vii, x.

11. Fritz Ringer, "The German Academic Community," in Oleson and Voss, p. 419.

12. David Noble, *America by Design: Science Technology, and the Rise of Corporate Capitalism* (New York: Oxford University Press, 1977), p. 43.

13. Donald M. Pattillo, *Pushing the Envelope: The American Aircraft Industry* (Ann Arbor: University of Michigan Press, 1998), p. 20.

14. See Peter Jakab, *Visions of a Flying Machine: The Wright Brothers and the Process of Invention* (Washington: Smithsonian Institution Press, 1990).

15. Pattillo, p. 29; Aeronautical Chamber of Commerce of America, Inc., *Aircraft Year Book for 1924* (New York: Aeronautical Chamber of Commerce of America, Inc, 1924), pp. 312–313.

16. Wilfred B. Shaw, ed., *The University of Michigan, an Encyclopedic Survey* (Ann Arbor: University of Michigan Press, 1941–), p. 1184.

17. Crouch, pp. 293–95; Roger Bilstein, "American Aerospace Technology: An International Heritage," in Galison and Roland, p. 208.

18. I.B. Holley, *Ideas and Weapons* (Washington: Air Force History and Museums Program, 1997), pp. 103, 106; University of Michigan, *The President's Report for the Year 1920–1921* (Ann Arbor: University of Michigan, 1922), p. 198.

19. U.S., National Advisory Committee for Aeronautics, *Annual Report, 1921* (Washington: Government Printing Office, 1922), p. 27.

20. U.S., National Advisory Committee for Aeronautics, *Training Aeronautical Engineers* (1922), by Edward Warner, NACA Technical Memorandum No. 149, 1922; originally published in the *Christian Science Monitor* on September 5, 1922. The NACA publications had very wide circulation.

21. Armour Institute of Technology, Chicago; California Institute of Technology; Cornell University; Indiana University; Massachusetts Institute of Technology; Purdue University; Rensselaer Polytechnic Institute; "Research University, Washington, DC"; Stanford University; State University of Iowa; University of California (Berkeley); University of Detroit; University of Illinois; University of Michigan; University of Wisconsin. *Aircraft Year Book 1922*, p. 230.

22. *Training Aeronautical Engineers* (1922), by Edward Warner, NACA Technical Memorandum No. 149, 1922, p. 1.

23. Ibid., pp. 1–2.

24. Ibid., pp. 3–4.

25. Ibid., pp. 4–5.

26. *Catalogue of the University of Michigan, 1916–1917* (Ann Arbor: University of Michigan, 1917), p. 379.

27. M. E. Cooley to G. Douglas Wardrop, 19 April 1916, University of Michigan College of Engineering Records, Box 8 (April 1916), Bentley Historical Library, University of Michigan, Ann Arbor.

28. University of Michigan, *The President's Report for the Year 1920–1921* (Ann Arbor: University of Michigan, 1922), p. 198; University of Michigan, *The President's Report for the Year 1921–1922* (Ann Arbor: University of Michigan, 1923), pp. 37–39; Shaw, *UMich Encyclopedia*, pp. 1183–84.

29. Reginald M. Cleveland, *America Fledges Wings: The History of the Daniel Guggenheim Fund for the Promotion of Aeronautics* (New York: Pitman Publishing Corp., 1942), pp. 1–2.

30. Richard P. Hallion, *Legacy of Flight: The Guggenheim Contribution to American Aviation* (Seattle: University of Washington Press, 1977), p. 30, quotes Harvey O'Connor, *The Guggenheims: The Making of an American Dynasty* (New York: Covici, 1937), p. 425.

31. Ultimately, the Fund would give more than $3 million away. The Fund's four priorities were: 1. To promote aeronautical education both in institutions of learning and among the general public; 2. To assist in the extension of fundamental aeronautical science. 3. To assist in the development of commercial aircraft and aircraft equipment. 4. To further the application of aircraft in business, industry, and other economic and social activities of the nation.

32. U.S., National Advisory Committee for Aeronautics, *Annual Report, 1926* (Washington: Government Printing Office, 1927), p. 68.

33. The grant amounts were as follows: California Institute of Technology = $305,000; University of Michigan = $78,000; Massachusetts Institute of Technology = $230,000; University of Washington = $290,000; Stanford University = $195,000; Georgia Institute of Technology = $300,000. From Cleveland, pp. 137–165.

34. *Aircraft Year Book 1926*, p. 301.

35. Pattillo, p. 74.

36. *Aircraft Year Book, 1926–1934.*

37. *Aircraft Year Book 1926*, pp. 97–98.

38. *Aircraft Year Book 1927*, pp. 116–117.

39. http://www.nasulgc.org/publications/Land_Grant/1890_Act.htm.

40. *Proceedings of the Forty-Third Annual Convention of the Association of Land-Grant Colleges and Universities*, Chicago, November 12–14, 1929, edited by Charles A. McCue, published April, 1930, pp. 444–445, Record Series 10/3/53, NASULGC Proceedings, Box 1, University of Illinois Archives, Urbana-Champaign, Illinois. Used with permission from the National Association of State Universitities and Land-Grant Colleges.

41. *Proceedings of the Forty-Fourth Annual Convention of the Association of Land-Grant Colleges and Universities*, Washington, DC, November 17–19, 1930, edited by Charles A. McCue, published April 1931, pp. 308–309, Record Series 10/3/53, NASULGC Proceedings, Box 1, University of Illinois Archives, Urbana-Champaign. Used with permission from the National Association of State Universitities and Land-Grant Colleges.

42. Ibid., pp. 317–318.

43. Ibid., p. 319.

44. *The Bulletin of the University of Minnesota, College of Engineering and Architecture and School of Chemistry, 1930–1931*, vol. 33, no. 30, July 5, 1930, p. 27, University Archives, University of Minnesota-Twin Cities, Minneapolis.

45. "The 1932 Alumni Directory," *The Minnesota Techno-Log* 12 (June 1932): 211–251, University Archives, University of Minnesota-Twin Cities, Minneapolis.

46. University of Minnesota Department of Aeronautical Engineering, *Fifty Years of Aeronautical Engineering: University of Minnesota, 1929 to 1979* (Department of Aeronautical Engineering, University of Minnesota, Minneapolis.), np., University Archives, University of Minnesota-Twin Cities, Minneapolis; *Aircraft Year Book 1928*, pp. 180–181.

47. Bulletin of the University of Minnesota, *Report of the President for the Biennium 1930–32*, vol. 35, no. 64, December 1, 1932, p. 141, University Archives, University of Minnesota-Twin Cities, Minneapolis.

48. The Bulletin of the University of Minnesota, College of Engineering and Architecture and School of Chemistry, 1930–1931, vol. 33, no. 30, July 5, 1930, p. 27,

University Archives, University of Minnesota-Twin Cities, Minneapolis; *Aircraft Year Book 1928*, p. 181.

49. William F. Durand, "Aerodynamic Laboratory of the Leland Stanford Junior University," *Journal of the Society of Automotive Engineers* 2 (March 1918): 230–238; Stanford University, *Annual Register*, vols. 1917/18–1940/41 (Stanford: Stanford University, 1917–1941); R. L. Wilbur, Stanford to The Trustees of the Daniel Guggenheim Fund, 13 May 1926, President R. L. Wilbur Papers, Mss Collection SC64A, Box 62, Fldr "Mechanical Engineering," Stanford University Archives, Stanford University, California.

50. R. L. Wilbur, Stanford to The Trustees of the Daniel Guggenheim Fund, 13 May 1926, President R. L. Wilbur Papers, Mss Collection SC64A, Box 62, Fldr "Mechanical Engineering," Stanford University Archives, Stanford University, California.

51. "Alfred Salem Niles" and "Elliott Gray Reid" in The Illinois Writers' Program of the Work Projects Administration, *Who's Who in Aviation, 1942–43* (Chicago: Ziff-Davis Publishing Company, 1942), pp. 313, 352; Elliott Reid to E. P. Lesley, 18 April 1927, Elliott G. Reid Papers, Mss Collection SC 230, Box 9, Folder 27, Stanford University Archives, Stanford University, California; William F. Durand to Theodore J. Hoover, Stanford University, 16 May 1927, President R. L. Wilbur Papers, Mss Collection SC64A, Box 64, Fldr "Eng-Mechanical (Guggenheim)," Stanford University Archives, Stanford University, California.

52. "Flying in a Tunnel," *Stanford Illustrated Review* (March 1930): 288–290.

53. A. S. Niles and J. S. Newell, *Aircraft Structures* (New York: John Wiley, 1929).

54. "Preface to the First Edition" in A. S. Niles and J. S. Newell, *Aircraft Structures*, vol. 1, 2nd ed. (New York: John Wiley & Sons, 1938), p. x.

55. *Proceedings of the Forty-Fifth Annual Convention of the Association of Land-Grant Colleges and Universities*, Chicago, Illinois, November 16–18, 1931, edited by Charles A. McCue, published April, 1932, pp. 504–506, Record Series 10/3/53, NASULGC Proceedings, Box 1, University of Illinois Archives, Urbana-Champaign. Used with permission from the National Association of State Universities and Land-Grant Colleges.

56. *Proceedings of the Forty-Sixth Annual Convention of the Association of Land-Grant Colleges and Universities*, Washington, DC, November 14–16, 1932, edited by Charles A. McCue, published March 1933, p. 481, Record Series 10/3/53, NASULGC Proceedings, Box 1, University of Illinois Archives, Urbana-Champaign. Used with permission from the National Association of State Universities and Land-Grant Colleges.

57. Montgomery Knight, "The Aeronautical Engineer," Radio talk, April 5, 1932, Aerospace Engineering Records Ms 172, Box 1.34, Archives, Library and Information Center, Georgia Institute of Technology, Atlanta.

58. Susan Butler, *East to the Dawn: The Life of Amelia Earhart* (Reading: Addison-Wesley, 1997), pp. 306–308, 310–11, 317.

59. Montgomery Knight, "Aeronautics and the Engineer," *Technique*, Friday, November 8, 1935, Aerospace Engineering Records Ms 172, Box 1.12, Archives, Library and Information Center, Georgia Institute of Technology, Atlanta.

60. Georgia School of Technology, *Bulletin of the Georgia School of Technology*, 28 (April 1931), pp. 25–29.

61. Knight's papers contain many examples. Here I drew upon an outline and manuscript ("The New School of Aeronautics at Georgia Tech") written for publication in the *Atlanta Journal* dated February 25, 1931, Aerospace Engineering Records Ms 172, Box 1.36, Archives, Library and Information Center, Georgia Institute of Technology, Atlanta.

62. Montgomery Knight to Clark B. Millikan, 1 July 1930, Clark B. Millikan Collection, Box 5.3, California Institute of Technology Archives, Pasadena.

63. A. L. Klein to Montgomery Knight, 24 July 1930, Clark B. Millikan Collection, Box 5.3, California Institute of Technology Archives, Pasadena.

64. Judith R. Goodstein, *Millikan's School: A History of The California Institute of Technology* (New York: W.W. Norton, 1991), pp. 156–163; Paul A. Hanle, *Bringing Aerodynamics to America* (Cambridge: MIT Press, 1982), pp. 12–18; Daniel J. Kevles, *The Physicists: The History of a Scientific Community in Modern America* (Cambridge: Harvard University Press, 1987), pp. 155–56.

65. Hanle, p. 17.

66. Michael H. Gorn, *The Universal Man: Theodore von Kármán's Life in Aeronautics* (Washington: Smithsonian Institution Press, 1992), chapter 3, "From Aachen to Pasadena," pp. 33–54.

67. Goodstein, pp. 169–173.

68. "Aero. Advanced Education," Speech delivered [by Clark Millikan] before Southern California NAA, November 25, 1929, Clark B. Millikan Collection, Box 17. 5, California Institute of Technology Archives, Pasadena.

69. *The Guggenheim Aeronautical Laboratory of the California Institute of Technology: The First Twenty-Five Years* (Pasadena: The California Institute of Technology, 1954), p. 7.

70. Clark B. Millikan to Montgomery Knight, 5 September 1939, Clark B. Millikan Collection, Box 5.3, California Institute of Technology Archives, Pasadena.

71. "The Daniel Guggenheim Graduate School of Aeronautics of the California Institute of Technology: A History of the First Ten Years," *Bulletin of the California Institute of Technology* 49(May 1940), pp. 5–6.

72. *Aircraft Year Book*, 1934–1941.

73. J. L. Atwood to Montgomery Knight, 12 October 1938, Aerospace Engineering Records, Ms 172, Box 1.14, Archives, Library and Information Center, Georgia Institute of Technology, Atlanta.

74. Ibid.

75. Montgomery Knight to ——, 1938. A copy of the form letter and a list of 21 companies this letter was sent to is included in Aerospace Engineering Records Ms 172, Box 1, Folders 13 and 14, Archives, Library and Information Center,

Georgia Institute of Technology, Atlanta. The companies include United Aircraft, American Airlines, Beech, Boeing, Chance Vought, various Curtiss-Wright Corporation subsidiaries, Douglas, Eastern Air Lines, Lockheed, Glenn L. Martin, North American Aviation, Pan American Airways, Piper, Pratt and Whitney, Transcontinental and Western Air, United Air Lines, Vultee, Western Air Express and Wright Aeronautical.

76. Montgomery Knight, "Aeronautical Education," A paper presented at the 1939 Annual Meeting of the Institute of the Aeronautical Sciences, Aerospace Engineering Records Ms 172, Box 1.13, Archives, Library and Information Center, Georgia Institute of Technology, Atlanta.

77. Shatswell Ober, "Discussion of Curricula for Aeronautical Engineering," December 9, 1938, Mss Collection AC 43, Box 17.29, Institute Archives, Massachusetts Institute of Technology, Cambridge.

78. C. T. Reid to Montgomery Knight, 24 January 1939, Archives and Records Management Department, Aerospace Engineering Records Ms 172, Box 1.14, Archives, Library and Information Center, Georgia Institute of Technology, Atlanta.

79. Shatswell Ober, "Discussion of Curricula for Aeronautical Engineering," December 9, 1938, Mss Collection AC 43, Box 17.29, Institute Archives, Massachusetts Institute of Technology, Cambridge.

80. Ibid.

81. J. W.Vale, Jr., to Montgomery Knight, 11 October 1938, Archives and Records Management Department, Aerospace Engineering Records Ms 172, Box 1.14, Archives, Library and Information Center, Georgia Institute of Technology, Atlanta.

■ Engineering National Defense

Technical Education at Land-Grant Institutions during World War II

Amy Sue Bix

Like other national institutions, most colleges in the United States had faced difficult times during the 1930s. Economic depression forced many students to drop out and strained school budgets to the limits. Yet again, like the rest of the country, international events toward decade's end distracted people on campus from domestic issues. Students and faculty closely observed the developing conflict in Europe, debating political and practical questions of American intervention.

During the debate over American preparedness, engineering education came to the forefront. It seemed obvious that as modern warfare came to rely on increasingly complex weaponry, countries must call on those with advanced technical skills and knowledge. Despite the image of academia as an ivory tower, a sense of national emergency hung over engineering programs as early as the fall of 1940. Especially at land-grant schools such as Iowa State, Penn State, and Cornell, where engineering had historically been a central component of the institution's educational mission, issues of America's military and industrial readiness acquired new importance. Once the United States entered World War II, military images filled engineering departments. Celebrating their annual "Engineers' Day" in 1942, Minnesota students portrayed themselves as "engineers going to bat for Uncle Sam." They produced special buttons showing an American eagle with its chest thrust out aggressively, a slide rule hanging prominently on its belt.[1]

After the attack on Pearl Harbor, the Society for the Promotion of Engineering Education issued a statement saying that America's declaration of war "undoubtedly requires an increase in the speed and extent of the preparedness program. The demand for engineers in both military and civilian service will grow correspondingly." The SPEE polled administrators of engineering colleges about whether academic schedules should be accelerated so that students could enter the workforce or military sooner. After studying responses from 125 schools, the SPEE concluded that "the present senior class in engineering [should] be graduated as early as possible consistent with the maintenance of a sound engineering educational program." It recommended that schools should add summer sessions so that juniors could graduate three months earlier and sophomores eight months ahead of time.[2]

Many land-grant colleges and other schools accordingly compressed academic calendars, shortening or eliminating normal vacations. Citing wartime urgency, Cornell required all engineering students to attend class three semesters per year, letting them finish four-year degrees in three years and five-year programs in four. New freshmen began class in summer, rather than waiting until September. Some administrators had protested that forcing engineering students to attend class year-round would deprive them of the summer earnings many counted on for tuition. When Cornell switched to an accelerated program, its trustees readjusted scholarships and loans to provide additional financial assistance.

In contrast to the 1930s, when graduating engineers feared they might not find work, Cornell's wartime students knew their future:

> In any typical group of engineering seniors, two out of three are enrolled in the advanced ROTC or in the US Naval Reserve, ready to go on active duty . . . immediately after Commencement. Of the 218 members of the Engineering Class of 1942, 83 will become second lieutenants in the Army; 38 have been commissioned in the Naval Reserve, and at least 25 more will be commissioned and ready for active service in the Navy before . . . June. Most of the rest will go to positions in industries or in engineering concerns working on war contracts.[3]

Further complicating the situation, college officials engaged in battle with local draft boards. Selective Service often refused to grant engineering students a deferment, despite the appeals of school administrators and engineering societies that pointed out that intelligent young men would serve their country better as trained engineers than in the ranks. By 1943, male civilian enrollment on campuses across the country plunged.

Engineering Training for the Military

Low civilian enrollment did not mean that engineering programs became idle. West Point alone could not turn out enough technical specialists for the military, so in 1942, the Army created plans to fill that shortage as soon as possible. The Army Specialized Training Program sent soldiers with superior educational backgrounds and test scores to colleges across the country, where they took prescribed courses in engineering, science, and math. The basic phase of ASTP condensed the first one-and-a-half years of college into nine months. Advanced ASTP gave soldiers accelerated training in aeronautical, chemical, civil, electrical, mechanical, and sanitary engineering, plus surveying, communications, marine transportation, acoustics and optics, and more. Although ASTP men lived on campus, they remained on active duty under military discipline and received regular army wages. Upon completing courses, ASTP men were assigned to technicians' and specialists' posts in the army, the Corps of Engineers, Chemical Warfare Service, Signal Corps, or Ordnance Department. By late 1943, the army had set up ASTP units at over two hundred colleges and universities. At Penn State, almost eight hundred men arrived at once.[4]

The navy similarly sent hundreds of men to Cornell, Penn State, North Carolina, Iowa State, and other colleges to study engineering, communication, electrical systems, and other technical fields. The U.S. Naval Training School at Cornell alone trained 2,001 officers in diesel engineering and 695 in marine steam engineering. The navy paid to build a new wing on Cornell's engineering lab and stocked it with almost every ship engine available. To teach navy men, Cornell brought men from engine manufacturing companies to join regular engineering faculty. Trainees gained both theoretical knowledge and hands-on expertise in diesel power, engine and hull construction, propeller design, and engineering physics. After finishing the four-month course, they were detailed as engineering officers at cargo and repair bases or on patrol vessels, mine sweepers, and other small craft. Navy public relations bragged that Cornell had transformed a "doctor of jurisprudence, just out of Harvard Law School . . . into an excellent engineer who is now at sea on a submarine chaser. A man with an LL.B. from the University of Chicago won honors in an examination in thermodynamics."[5]

The federal government provided a number of other programs through which military personnel entered wartime engineering studies. Purdue, for example, gave air corps engineering officers a three-month training course in aircraft maintenance. Military men arrived in such numbers and so fast that it stretched campus facilities to the limit. Cornell and Purdue converted dorms, fraternity houses, and even faculty residences into army and navy housing, squeezing in extra cots to accommodate still more. Civilian students were pushed into off-campus housing. At Iowa State, by September 1943, students in the armed forces outnumbered civilians by more than seven hundred. Civilian enrollment had fallen by more than 1,000 since spring, but the arrival of 2,314 navy men, 876 army men, and 91 special female engineering students kept total enrollment almost normal. Iowa State observers marveled at the visible changes wartime wrought on the engineering campus:

> New London and San Diego were names [formerly] connected with technical training for . . . the Navy—not Ames, Iowa! But . . . Iowa State has become one of the largest college Navy bases in the country. . . . V-12's pack engineering library and classrooms during all hours. . . . Naval Air Cadets have classes at night, burning the midnight oil to learn more about navigation and codes. Platoons of Army men are seen marching to and from mess. . . . Instead of the usual eight or ten sections of a pre-war statics class, civilian students are limited to two or three.[6]

The campus paper editorialized with pride that war programs brought new recognition to Iowa State engineering, underlining the growing importance of technical education in the twentieth century. "Looking to the future, we . . . are told that peace will bring . . . new technology—greater and more comprehensive than anything the world has yet known or imagined. Believing this is to believe in a brighter future for the nation—and for the technical school."[7]

Engineering Training for Defense Workers

As substantial as the programs to give military personnel technical training on college campuses were, government efforts to spread engineering education also extended to almost two million civilians. In the spring of 1940, with the continued advance of Nazi forces, the U.S. Office of Education began considering what part it might play in a future military emergency. Congress soon approved a program to extend *vocational* training, but John Studebaker, Commissioner of Education, foresaw a deeper problem in higher education. He worried about a straightforward crisis of supply and demand; colleges simply weren't graduating enough engineers to fill personnel shortages which already appeared as defense-related industries expanded production. Studebaker asked Andrey A. Potter, Purdue's dean of engineering, to join his staff as a consultant. Potter wholeheartedly embraced the idea of expanding federally-funded training to college engineering, writing, "Irrespective of the outcome of the war in Europe, there is bound to be in the near future keen industrial competition. Our country must meet this competition by more scientific and technological knowledge. . . . Higher education, particularly in science and technology, is also a major essential in our military defense . . . [which] depends upon ships, airplanes, tanks, gunpowder, and other manufactured products."[8]

The Office of Education launched surveys in New York City, Chicago, California, and Pennsylvania to gauge regional personnel needs. After conferring with company managers, educators, and engineering societies in more than forty communities, Penn State's extension service concluded that the state needed 7,500 new technical specialists, mainly qualified engineers. Studebaker summoned presidents and deans from the nation's leading engineering colleges to Washington, to meet representatives from the army, navy, and Office of Education. To their dismay, educators heard that the United States already faced "a marked shortage in naval architects, ship draftsmen, marine engineers, engineers skilled in airplane structures, airplane power plants and airplane instruments."[9]

Technical knowledge represented a limited national resource. Lehigh president C. C. Williams wrote, "Colleges have no synthetic chemistry in sight that will make engineers out of air and coal." But though schools could not instantly expand their classes of graduating engineers, they began thinking about ways to supplement regular undergraduate and graduate programs. The hope was to create specialized crash training which could eliminate potential personnel bottlenecks in critical defense industries. Drawing up initial plans, the Office of Education calculated that during the present year alone, the government, the navy, and the aviation industry could use at least 2,500 more people familiar with airplane structures, power plants, and stress analysis. Airplane and ship builders could also absorb at least four thousand engineering draftsmen, according to the Civil Service Commission. Studebaker and Potter figured that engineering freshmen or high-school graduates with shop experience could easily pick up mechanical drawing, freehand detailing, and structural analysis in twelve or sixteen weeks. If one hundred colleges each put together one such class, the nation would have five thousand additional draftsmen available within four months. Experts also worried particularly about a shortage of ship draftsmen. Normally, only three schools (MIT, Michigan, and Webb Institute) even taught naval architecture. Those places had graduated just fifty-one students in 1940, but within two years, the navy and shipbuilders might well need four thousand marine engineers. To supply that brainpower, educators suggested retooling civil, mechanical, electrical, or architectural engineers through a twelve-week course in ship geometry, theory of navigation, marine engineering, and special design problems.[10]

In October 1940, Congress passed a bill along lines recommended by Studebaker and Potter, appropriating nine million dollars to establish the new Engineering Defense Training program, based in the U.S. Office of Education. That legislation defined EDT's mission as "providing short intensive courses on the engineering college level in fields essential to national defense . . . where a shortage of trained personnel prevails at present or is almost certain to occur if steps are not soon taken." Potter, a former president of the SPEE, the American Society of Mechanical Engineers, and the American Engineering Council, would chair EDT's national advisory committee of engineering educators. Dean R. A. Seaton of Kansas State came to Washington to direct day-to-day administration.

Studebaker knew that EDT would not succeed without the public commitment of major technical institutions. Accordingly, he appointed prominent engineering deans (including H. P. Hammond from Penn State and Gill Gilchrist of Texas A&M) to serve as national advisors. Deans from Cornell, Ohio State, Michigan State, Kansas State, Maryland, and the University of Texas became regional advisors, acting as liaisons between the U.S. Office of Education and their region's engineering colleges and defense industries. Those representatives immediately brought their own schools on board and appealed to colleagues at neighboring institutions to sign on. Any institution with accredited engineering curricula could sponsor EDT courses, and most were pressured to do so by Washington headquarters and the SPEE. The new program would be a "cooperative effort in which the federal government furnishe[s] the funds and the colleges furnish the working . . . facilities."[11]

Under the system set up, college officials were expected to identify existing and anticipated personnel shortages in local defense industries, then propose ways of meeting those needs. Schools were to plan both full-time and part-time courses, each "designed to train for a specific activity of *immediate* application." Full-time ("pre-employment") courses were intended to teach unemployed men with some technical ability to become engineering assistants, technicians, and draftsmen. RPI president William Hotchkiss explained, "There is a large reservoir of persons . . . partly trained, . . . who have had science courses in liberal arts colleges, or who have had to drop out of engineering courses before graduation, who could be quickly prepared for service in a few weeks or months." EDT also hoped to retrain men who already held engineering degrees to handle new fields. "A civil engineer, through a four-months intensive course, may be prepared to design airplane structures, or a mechanical engineer . . . qualified for airplane engine design." Part-time ("in-service") courses would provide after-hours training, upgrading men already in defense work to positions of more responsibility and improving their technique.[12]

Regulations stipulated that EDT courses must run at "college-grade," which meant anything from freshman level to graduate work. Participants would be required to hold a high school diploma; advanced courses might specify two years of college or even a degree as prerequisites. Courses would be entirely free to qualified students, with the government paying all tuition and lab fees. Seaton stressed that "[c]ompletion of an EDT course should prepare a trainee for immediate employment in a defense activity, if he is not already so employed, and he should be immediately available for such employment after he completes the course."[13]

Isador Lubin, commissioner of the Bureau of Labor Statistics, warned the EDT office that defense expansion threatened to overwhelm company personnel offices. The government had already awarded more than eight billion dollars' worth of defense contracts, entailing an extra four and half million man-years of labor. "New products not hitherto manufactured in this country require new skills that must be developed by training. In normal industry only 20% of workers are skilled; in defense industries this will rise to 40% to 60%." EDT officials realized they must "anticipate the needs of industry [since] it will take three or six months before our trainees are ready. It isn't enough to wait until the need is immediate." Harassed superintendents tended to wait until the last minute, then search colleges for "nice clean-cut-looking [engineering] lads with flat feet or dependents, who won't be drafted."[14]

Training efforts started off slowly in areas distant from defense activity, such as Iowa, Arizona, and Arkansas. The University of New Mexico, Idaho, and Montana State College attempted to organize drafting classes, as requested by national advisors, but failed to secure enough qualified students. With rapid expansion of airplane and ship building, however, California, Washington, and the Northeast saw immediate demand for well-trained employees. By the end of 1940, the national office had approved 418 courses at 84 institutions for more than 29,000 trainees. Most courses met two or three evenings per week, generally for twelve to sixteen weeks. Subjects most in demand included production supervision, with 6,641 students; engineering drawing, with 4,614 students; production engineering, with 3,364; materials inspection and testing, with 3,152; metallurgy, with 2,646; tool engineering, with 2,059; and machine design, with 1,889. EDT staff feared that initial enrollments still fell "far short of meeting requirements, especially in ... aeronautical engineering, explosives, machine design, naval architecture and marine engineering." Washington encouraged colleges to add more such courses, since plans for expanding the nation's shipyards were underway, and "we must not allow the shortage of engineers to delay this program."[15]

SPEE president Donald Prentice brought out statistics to document this manpower gap. Estimates suggested that government and industry would need 40,000 or 50,000 new engineers in 1941, yet only about 12,000 students were on track to complete engineering degrees in June. Moreover, one-third of graduates would not be available for civilian employment; approximately 4,000 would be given ROTC commissions, drafted, or recruited as naval reserve officers and pilots. Penn State reported that 1941 was already "the most hectic recruiting season we have ever experienced." Major corporations were "hoping to secure not only their usual quota of graduates, but a much enlarged one," while "there will also be many requests from industries who haven't made requests for a long while and some who have never before." Eager employers had begun a bidding war, sending up engineers' salaries considerably. "It is evident that there is going to be a very wide gap between the supply of young engineers and the demand for them," Dean Hammond concluded. "Twice as many wouldn't be too many." On the opposite coast, placement officers at the University of California declared, "We cannot find even one engineer to refer to the dozens of employers who are clamoring for them.... [I]n sheer desperation employers come personally to explain their needs. One firm not in existence six months ago has 200 employees today and tells us they will need 2000 within another month."[16]

The Office of Education publicized its new program extensively. An NBC radio show entitled "Engineers for Defense" described training opportunities to a nationwide audience. Colleges broadcast announcements of their own classes, advertised in local papers, and sent information booklets and posters to the area's major employers. In some cases, a flood of subsequent inquiries overwhelmed coordinators; the University of Detroit fielded 1,500 requests for enrollment in its first courses.

By June 1941, Washington had approved proposals for 2,350 courses at 144 schools, reaching 136,618 students. Building on the administrative infrastructure already in place for its regular extension program, Penn State alone opened EDT courses for over 10,000 workers in fifty cities in response to local companies' requests for help. When Piper Aircraft expressed a need for trained draftsmen and junior engineers, Penn State set up classes in aircraft layout, operations inspection, aircraft structure, metallurgy of welding, aircraft drafting, time and motion study, fundamentals of engineering, and aerodynamics. By 1944, 65 percent of the personnel in Piper's engineering department had completed at least one course through Penn State. A supervisor-mechanic who took engineering drafting had been promoted to chief design engineer in the experimental department. A former shoe clerk in Piper's sewing department became assistant chief draftsman. A shop employee had risen to work on plant layout and scheduling efficiency in Piper's new methods engineering department.[17]

Arranging EDT courses consumed substantial time and effort for school administrators. By the spring of 1941, the University of Florida was devoting more resources to EDT than to its regular engineering program. The University of Cincinnati stated its willingness to double EDT offerings, but warned that finding good teachers for extra classes could prove difficult since "our university teaching staff is taxed to the utmost." On balance, D. V. Terrell of the University of Kentucky spoke for many when he observed that after the first round of EDT training, "[i]t is sometimes difficult to say just where the national defense program is benefiting. However, it may be said that such education will be reflected in future needs for men trained to go direct into the defense industries, or to fill gaps left by those who do go into such industries."[18]

EDT courses soon won praise from companies such as Lockheed, which had at least 575 workers enrolled in courses run through Caltech. "This program has filled a need which we have long felt but about which we were able to do little because of the expense involved in . . . regular university extension classes." Cadillac's staff observed that automakers were being asked to turn out new products for which "the tolerances are more exacting and the inspection more rigid than in our normal manufacture. We will have to do training ourselves, but the more foundation we have on which to build, the better off we will be. Training in engineering colleges and the EDT should be increased . . . to relieve industry of this burden."[19]

Some firms reported that having workers participate in EDT classes had already yielded direct benefits in production. At the American Can Company, workers learning time and motion study had pointed out sources of waste in routing materials and suggested ways to improve floor plans. The availability of training had convinced the First Sterling Steel Company to adopt new processes. As the University of Pittsburgh reported, "An employee . . . taking the course in x-ray testing brought the director of research as a visitor. . . . As a result, the company is planning to buy equipment for radiographic examination of their products."[20]

EDT students themselves appreciated EDT as "a common-sense safeguard" for military readiness. One man taking a photogrammetry class wrote, "This subject is fast becoming important in defense mapping and there are few up to date text books available. I am becoming skilled in the use of laboratory equipment which would be impossible to see or use in any other way. I registered . . . with the thought . . . that engineers might soon be conscripted for all defense measures, and the training I am receiving will enable me to do my share in this important phase of work." Another explained, "I took the course in water supply and sanitation because I had time to spare and it never hurts to keep brushed up.... Courses . . . keep the 'white collar' man in condition the same way the army camp keeps the soldier in condition."[21]

Participants also anticipated personal benefit from training. Although family obligations or economic constraints might have prevented workers from seeking education in peacetime, EDT training could finally qualify them for promotion. One Michigan student commented, "Being a tool and die maker that came up the hard way with no education . . . your courses have kindled a fire in me for more.... I was a die maker from the old school and with the aid of your course I am now able to go ahead." One goal of the U.S. Office of Education had been to give jobless people engineering training which could help them land employment. While the program's overall success along those lines was difficult to rate, Penn State reported that almost two thousand students who were unemployed upon beginning class had since found positions. One still out of work felt "sure that this course is adding much toward placing me in the near future.... [It] gives me the feeling that I am one of the very few learning an entirely new branch of engineering."[22]

Although EDT had been created and implemented by leading engineering educators, some inside that community remained ambivalent. Critics worried about setting a dangerous precedent which could give the government an excuse to control engineering education. If America entered the war, they feared, the emergency might well bring an "insistence that the army run everything." The military might be tempted to convert colleges into something resembling West Point and Annapolis, which would be "horrible . . . for the training of civilian technical men."[23]

Another threat soon appeared: hoping to get in on the action, junior colleges, technical institutes, liberal arts colleges, and schools of commerce lobbied Congress to let them offer courses. The prospect of turning training over to outsiders horrified EDT advisors, who considered many proprietary schools illegitimate and pointed out that most junior colleges didn't even have engineers on their faculty. EDT's board passed a motion declaring that "in order to protect professional engineering training," standards and procedures must "remain in the hands of the engineering colleges."[24]

Lehigh's C. C. Williams warned further that over-expansion of EDT created "a danger of 'inflation' in engineering education." He declared, "To stamp a large number of men as having had engineering training who are not trained engineers [is like] . . . running the printing presses to turn out paper currency.... We are in danger of depreciating or destroying the value of the real thing." Williams and others feared that trainees might consider themselves "graduates" of Cornell or Penn State after taking one or two EDT courses through those schools. EDT organizers emphasized that the program did not teach men with a liberal arts background to "become engineers in three easy lessons."

Though EDT was run through the nation's leading engineering departments, most institutions did not give participants credit toward degrees. After all, EDT courses were "not given as integrated parts of an [engineering] education." Training classes were expected to reflect "an academic standard customarily required of college and university students in the same field"; however, the approach was not identical. EDT studies were "[o]rdinarily ... of a more intensive and applied character in order to give specific training for a particular field of war work." The Engineers' Council for Professional Development warned that if instructors began modifying specialized course content to satisfy demands for academic credit, it would interfere with EDT's primary objective, preparation for national defense. For that reason, both the ECPD and EDT's regional advisors passed resolutions recommending that colleges refrain from giving credit for war training. Instead, schools such as Texas A&M issued participants a certificate and wallet card, testifying that they had completed courses in airplane drafting, industrial safety, or other subjects.[25]

Despite such misgivings, engineering programs gained tangible advantages from participating in EDT. At a time when both industrial recruitment and the draft threatened to steal away college staff, EDT justified keeping faculty intact. Educators argued that any attempt to prepare technical workers should come through them, saying, "Just as the medical school is responsible for the training of medical technicians, so must the engineering school assume responsibility for the training of those who are to assist engineers." Furthermore, in addition to paying all course costs, the federal government gave schools an additional twenty percent to cover overhead expenses and purchase extra teaching supplies. Program rules allowed schools to keep new equipment, so that government money helped expand college labs in fields such as electronics. Ultimately, growing evidence of a nationwide manpower crisis overpowered lingering doubts. One company president noted, "It is almost impossible to get ready made engineers and it is going to be increasingly more difficult."[26]

Although initial legislation limited EDT's jurisdiction to engineering, that boundary proved hard to maintain. Classes in chemical engineering shaded near pure chemistry; students of airplane design needed to learn meteorology. Companies wanted supervisors to learn business administration, industrial methods, and accounting so that they could cope with the problems created by rapidly expanding production. Moreover, regulations stipulating that only engineering programs could run courses made it harder for schools to find teachers. The engineering dean at Virginia Polytechnic complained, "Some liberality in interpretation of *engineering* might be helpful. We might make more use of departments of physics and chemistry if we were not afraid to."[27]

NYU engineering dean Thorndike Saville worried that legislators would force engineers to cede control of EDT. "[I]t will be a great mistake to open up the gates to schools of commerce and business," he warned. Though business schools taught courses titled industrial engineering, the substance differed radically since "engineers have a background of a totally different experience from the men in the schools of commerce." Saville continued:

> Dean Seaton intimated that a good deal of pressure has been applied to have the law permit business schools . . . to participate in this program. . . . [I]t is just this matter of political pressure which has concerned a great many of the engineering and other college administrators with respect to the Office of Education programs. Many of us have contended from the outset that it will be impossible to avoid political pressures as the federal government more and more gets into education administration on the college level. This is a direct evidence that we were correct.[28]

Despite such criticism, Congress appropriated seventeen million dollars in July 1941 to add training in chemistry, physics, and production supervision to the program accordingly renamed "Engineering, Science, and Management Defense Training" (ESMDT).

Japan's December 1941 attack on Pearl Harbor lent new urgency to defense training. The Office of Education called for "universal war-mindedness," declaring, "Employed people . . . must be constantly trained for greater responsibilities." Its program changed names one last time, becoming "Engineering, Science, and Management War Training" (ESMWT). The Office of Production Management told engineering and science programs, "We need everything you can give us, as of yesterday."[29]

By now, training extended nationwide. Penn State had established the single largest program; in 1941–42, it enrolled almost 55,000 students in 150 different cities, one-eighth the national total. Since some locations did not have good teaching facilities, Penn State created three "auto-labs," trucks filled with scientific equipment. Those "chemistry and physics labs on wheels" traveled from town to town, demonstrating principles of electricity, mechanics, and matter to introductory science and engineering pupils. The full list of ESMWT courses taught through Penn State included:

> Advanced Inspection Methods, Aerodynamics, Air Conditioning, Aircraft Engines, Analysis of Solid and Gaseous Fuels, Applied Engineering Math, Applied Mechanics, Auditing, Ceramic Engineering, Chemical Thermodynamics, Chemistry of Engineering Materials, Chemistry of Explosives, Chemistry of Metals and Alloys, Coal Distillation, Combustion of Liquid and Gaseous Fuels, Corporation and Manufacturing Accounting, Cost Accounting for War Production, Cost Control, Die Design, Electric Meters and Instruments, Electric Motor Control, Elements of Radio Communication, Engineering Drafting, Foremanship, Foundations of Engineering, Fundamentals of Railway Signaling, Glass Technology, Heat Treatment for Tool Engineers, Heat Treatment of Steel I, II, and III, Industrial Electrics, Metallographic Laboratory Technique, Methods Engineer-

ing, Motor Freight Management, Industrial Supervision and Inspection for Women, Ordnance Inspection, Personnel Management, Petroleum Laboratory Technique, Physical Testing of Materials, Plastics Design and Machining, Production Control, Pyrometry, Qualifying Math for Engineering Courses, Safety Engineering, Shaping of Steel, Steam Power Plants, Surveying and Mapping, Time and Motion Study, Tool Design I and II, and more.[30]

Some subjects taught in ESMWT represented elementary material. When the York company requested help in finding drafting personnel, Penn State recruited twenty-three girls just out of high school for a full-time, 100-hour engineering fundamentals course. Twenty subsequently took posts as junior draftsmen at York. Ultimately, Penn State took 1,945 young people through introductory engineering, and 1,164 reportedly found work immediately. Typically, classes in "foundations of engineering" combined freshman-level math and physics with use of the slide rule and other basic engineering methods. The course description for prospective students explained, "Make no mistake. This course isn't a quick, easy way to become an expert engineer. It provides . . . powerful instruments of immediate practical utility to help you do a better job in industry. . . . What you build on it in the future is up to you."[31] Other training approached graduate-level science and engineering. Caltech's classes for people in petroleum refining introduced sophisticated principles and the use of infrared and ultra-iolet spectrophotometers.

Many ESMWT courses set out to fill specific manpower gaps. With schedules calling for production of smokeless powder alone to rise from fifty to one thousand tons per day, the army and industry desperately needed skilled explosives inspectors. Few colleges could hold training classes in explosives, since they did not have professors familiar with the subject. Accordingly, the Office of Education held special training for organic chemistry faculty from thirty-three institutions, who returned home to organize local courses on the chemistry of powder. One subsequently noted, "We never get an opportunity to complete a class," since arsenals and munitions companies "take them away from us before they finish."[32]

Building on that success, the Office of Education organized similar efforts to train experts in new radar technology. Electrical engineering and physics professors from forty schools met with army and navy officials in 1941 to outline a common course covering cathode ray tubes, amplification, oscillation, modulation and demodulation, receivers and transmitters, radiation, and more. After taking an intensive refresher course themselves at MIT, instructors brought training in ultra-high-frequency theory and methods to institutions such as Penn State. After Pearl Harbor, Iowa State required all electrical engineering seniors to take the two-quarter pre-radar course (waiving normal ESMWT rules to give them full degree credit). For teaching purposes, the government sent colleges thousands of dollars' worth of very special equipment.[33]

As the war accelerated, defense manufacturers were under great pressure. Large firms especially demanded ESMWT classes tailored to their exact work and limited to their own employees. Initially, the national advisory committee resisted; engineering educators felt that companies with such specific expectations should handle training themselves. Yet the urgency was apparent, and managers at companies such as Bethlehem Steel reported seeing the most satisfactory results from courses which industry people had helped plan. Industries repeatedly urged "extreme caution to prevent these courses from becoming too academic." Confirming such opinions, a study of Connecticut's entire EDT program revealed that different schools achieved varying success rates (defined as the percentage of trainees subsequently able to assume greater responsibility at work). The college with the highest proportion of successful students was the one with the closest coordination with the businesses being served.[34]

Even when educators feared losing academic rigor, they faced the hard fact that college faculty simply could not handle all the additional ESMWT courses on top of their regular teaching load. Out of both necessity and desire, many ESMWT courses were taught by men from industry (under supervision of academics). At Penn State, 83 percent of instructors were working engineers, and as one plant superintendent explained, "We like these courses ... especially because the instructors are practical men from industry."[35]

Both companies and the government boasted about links between employee training and improved performance. One Indiana machine manufacturer wrote, "We had trouble in getting good manganese bronze castings so we began making them ourselves. We [sent] test bars ... with our men when they went to the [metallurgy] class, ... and we are now making better castings than we can get anywhere else." West Coast airplane manufacturers claimed that sending people to courses on "quality control by statistical methods" had saved $800 on every bomber. The Rustless Iron and Steel Corporation staffed its new spectrographic laboratory entirely with employees who received their training through the program; their introduction of new techniques sped up analysis and thereby saved critical metals.[36]

Since the national office reviewed course offerings each year, the program maintained valuable flexibility. Toward war's end, changing defense priorities called for fewer courses on explosives and more in plastics, synthetic rubber, and petroleum refining. Colleges focused on serving regional needs. The University of Washington emphasized naval architecture, and USC and the Illinois Institute of Technology organized courses in the chemistry of food dehydration. Serving the state's oil industry, Penn State set up courses in petroleum laboratory techniques to train badly-needed technicians. One such class placed four unemployed women, two former secretaries, and one ex-salesclerk in the Pennzoil labs; another retrained two female soda fountain operators as core analysts.

The best example of how ESMWT served regional wartime demands came on the West Coast. The University of California ran the most courses of any institution, about one thousand, many targeted to aid airplane manufacture. School administrators explained, "At the present time the aircraft industries in the Los

Angeles area have great numbers of persons in the engineering departments who are capable only of the most elementary types of engineering activity. Approximately half of the people in this lower level must be upgraded to the second level. The places they vacate must then be filled by persons ... from the outside." To help upgrade workers, the university taught aircraft lofting, aircraft plastics, metals in aircraft design, aircraft development layout, stress analysis for aircraft designers, flutter prevention in aircraft, aerodynamics for designers, aircraft weight control, electrical engineering for aircraft, and aircraft industry administration. Enrollment in most of those relatively advanced courses required at least six months' experience in airplane manufacture. In order to bring new workers into industry, the university ran full-time drafting classes. After one such course in 1942, at least 107 out of 132 students (mostly female) found war-related work, 70 percent with either the Douglas, Ryan, Consolidated, or Vega airplane companies.[37]

California ESMWT staff had no reservations about getting too close to business. The university developed sixty-six classes for Lockheed staff at that company's request; Lockheed supplied both teachers and instructional material. Similarly, the Jacobs Aircraft Engine Company wrote, "The fact that you have worked with us in setting up a special course built around our particular engine ... [makes it] more effective." The University of Southern California put together twenty-five sections of a tool engineering course to meet at two o'clock in the morning for swing-shift employees. After gasoline rationing made it hard for workers to travel to campus, ESMWT moved courses directly into airplane plants. More than nine hundred workers at Consolidated-Vultee went through an in-plant ESMWT course on aircraft production control. "All students in the class are studying with the same frame of reference, their training needs are more narrowly definable, and illustrative examples are readily available in their common experience."[38]

Given rapid changes in airplane design, companies appreciated courses to help employees keep on top of new ideas. Curtiss-Wright found particular value in advanced training covering problems of airplane flutter and vibration. One aeronautical engineer requested a class on airfoils, since "this subject is changing so fast that not only our new employees but also our older ones not in direct contact with this work do not appreciate the present status."[39]

Initially, companies feared employees might not take ESMWT training seriously, since courses were free. Just the opposite; 60 or 70 percent of trainees completed class, attending regularly despite demands of late-shift employment and overtime. Virginia Polytechnic's engineering dean observed, "I never saw as serious a group of students in all my years." Students felt proud that new training helped them speed up war production. One wrote, "As a layout draftsman in the wing group of Consolidated's Engineering Department ... , a day hasn't past [sic] that I haven't applied analytic solutions [learned in ESMWT] to layout [sic] problems that would otherwise have been a nightmare." A former history major, who used a course on circuits and electric machinery to land a job at Boeing, reported,

"It's a grand feeling to be serving one's country in a technical way." Some found new knowledge exciting for its own sake. One wrote, "Before, . . . my job meant nothing to me but monotonous routine. Now with a better understanding of what happens [in] heat-treating, I find my work much more interesting and my superiors have noticed it. I am anxious to delve more deeply, working toward being a metallurgist."[40]

As the war began winding down, ESMWT ended in the summer of 1945. All told, 227 colleges and universities in every state (plus Alaska, Hawaii, and Puerto Rico) had offered courses. At a cost of about 94 million dollars, the federal government had paid to train 1,795,716 American men and women. More than 1.3 million had enrolled in various types of engineering: 207,618 had studied general engineering, 161,460 aeronautical engineering, 52,020 chemical engineering, 70,908 civil engineering, 250,563 electrical engineering, 184,977 industrial engineering, 43,762 marine engineering, 175,596 mechanical engineering, and 64,859 metallurgical engineering. Penn State alone had taught more than 140,000 people in 224 communities. Cornell trained 30,144 war workers through classes in Ithaca, Binghamton, Buffalo, Corning, Elmira, Endicott, Niagara Falls, and sixteen other locations.

Although ESMWT and its predecessors were the creations of wartime, their impact on American engineering lasted far longer.

ESMWT represented a unique episode, the most intense centralization of science and engineering education effort up to that time. With the United States facing an international crisis, university administrators grabbed at government support for extending and promoting engineering and science training. That immense commitment of public funds and professional effort helped pave the way for postwar policy developments. Although important differences exist, ESMWT in many ways foreshadowed the Cold War's National Defense Education Act. The 1950s once again brought announcements that the United States was not educating sufficient numbers of engineers and scientists. Experts warned that lack of brainpower could seriously compromise the race against communism. Extension education again took on new importance; in a 1956 speech to the American Institute of Electrical Engineers, U.S. Steel chairman Roger Blough praised evening classes as a means of upgrading shopfloor workers and technicians to great responsibility, as engineers. "The whole work force moves forward a notch or two, and that is another way to take care of an ever-increasingly mechanized America."[41] ESMWT had underlined the significance of engineering training to modern America's military readiness and industrial strength. It brought the university closer to the government and engineering programs closer to the military, trends which would only continue over the decades.

Engineering Training for Women

Wartime programs carried an even greater significance in the way they affected women's access to engineering training. For decades, formal barriers had maintained engineering as a primarily male preserve. Up to World War II and beyond, some of the nation's foremost technically oriented institutions (such as RPI, Georgia Tech, and Caltech) refused to enroll female undergraduates. Many male students, faculty, officials, and alumni at those elite schools openly criticized or ridiculed the idea of women engineers. Unwritten rules also discouraged women from attempting engineering education. The few coeds admitted to MIT struggled against a hostile intellectual and social environment. Long-standing school traditions tied technical expertise to masculinity.[42] World War II marked a brief break in such conditions, a window of opportunity which challenged assumptions about gender and engineering. Although at first glance it seems as if that window of opportunity closed all too suddenly at war's end, that temporary break paved the way for a long-term redefinition of women's place in college engineering departments.

Back in the late 1800s and early 1900s, a handful of women had established a small but significant place in engineering studies, attending mainly land-grant colleges or small technical schools. For instance, Olive Dennis received a civil engineering degree in 1920 from Cornell, then worked for over thirty years at the Baltimore and Ohio Railroad. Women such as Dennis got a certain amount of attention, since they were a rarity, a curiosity. A 1920s article remarking on that female presence was headlined, "Three Coeds Invade Engineering Courses and Compete With Men at Cornell University: Stand Well in Their Studies." Similarly, a 1925 article in the University of Minnesota's engineering publication bore the title, "Coed Engineers: Man's Domains are Again Invaded."

There was plenty of joking about women attempting to penetrate masculine territory. In 1938, Justin DuPrat White, vice chairman of Cornell's board of trustees, told the school's Society of Engineers, "When I entered Cornell in 1886 . . . two years before, . . . there had been a woman who had had the temerity to register as an undergraduate in Sibley College. I want to congratulate . . . the College of Engineering for their great tolerance, in these days of intolerance throughout the world, [in welcoming] the entrance of women. . . . Lord knows that the lawyers have almost been swamped with the influx of women in their profession." In fact, by 1938, more than twenty women had received engineering degrees from Cornell, one by one over the years. Isolation made the experience hard. One such "Slide Rule Sadie" (as they were nicknamed) said:

> A girl has to want . . . pretty badly to go through with the course in spite of the unconscious brutality of the young men who will be her classmates. . . . She must be ready to be misunderstood, as . . . many . . . will conclude that she took engineering . . . to catch a husband. She must be ready to do alone the work the men do in groups . . . lab reports, etc., because in general men who are willing

to face the scorn of their peers and . . . work with her are far more interested in flirting than in checking computations. She must be prepared for a pretty lonely academic career. . . . [43]

Before World War II, simply being a woman studying engineering was unique enough to get your picture on the front page of campus papers at Iowa State, Minnesota, and elsewhere. Those discussions of female engineering majors treated each one individually, as if each case were unusual—which it was. For instance, under a typically cutesy headline, "Beauty Meets Resistance," the *Penn State Engineer* noted that Olga Smith had become the school's first woman enrolled in electrical engineering. Yet beneath the joking, women's presence stirred controversy. In 1935, the *Iowa Engineer* commented:

> Things have reached a pretty pass when the girls can come over to the engineering side of the campus and beat the boys at what is theoretically their own game. Somehow it just doesn't seem to be quite right. But after all, honor to those to whom honor is due. At the Fall Honors Day Convocation, . . . a girl . . . walked off with the show. Alice Churchill, E.E. junior, received both the Pi Mu Epsilon calculus award and the Phi Kappa Phi high scholarship award. Perhaps this will serve as a stimulant to . . . embryo engineers of the opposite sex . . . to exert themselves a little more to uphold their much boasted "superiority" in such matters as engineering.[44]

Female engineering students themselves talked about their experiences in the singular—they simply didn't know enough others to refer to themselves as a group. Charlotte Bennett, who studied chemical engineering at Purdue in the 1930s, wrote, "I have surprised a good many people who thought I could not stick it out . . . [and] I would make the same choice" again.[45]

That image of women as solitary invaders venturing onto masculine ground came to the forefront with World War II, when there simply weren't enough male engineers available. Just as manufacturers turned to "Rosie the Riveter" on the shop floor, companies sought to begin employing women at drawing boards and in engineering shops. But of course, managers soon encountered the obvious problem—they couldn't find enough women with relevant training ready to move into engineering positions.

War provided a rationalization for giving more women access to engineering education. Purdue civil engineering student Ellen Ziegler wrote in 1942, "Think of the vast reserve of engineers we might have if we had been training women . . . during the past few years." The idea of women studying technical subjects suddenly acquired patriotic value. One month after Pearl Harbor, the University of Texas newspaper ran a photo showing the school's five female engineering majors working slide rules. The caption read, "No knitting or other sissy stuff for these five girls—they're doing their bit for national defense in a manly way."[46]

University announcements and campus publications were filled with bulletins urging women to take up engineering and science. Purdue's paper bemoaned reports that fewer than thirty percent of 1942's female college graduates had majored in fields needed for national defense. "The slide rule is only part of engineering, but those who can handle one well are needed to fill the vacancies left by the draft....A yen to build bridges or to know why an engine goes round is even more useful today than yesterday."[47]

To demonstrate to coeds how much potential they had, Purdue required all 1,300 female students to take a math and science aptitude test in December 1942. That announcement spread dismay across campus, with rumors that only women who scored well would be permitted to reenroll in spring. Administrators hastened to assure coeds that individual performances would not be held against them. Results were being compiled for national policy, "to find the percentage of college women who could be depended upon to replace men in jobs requiring some technical training." Meanwhile, Purdue handed out material promoting its science and math courses, emphasizing that all coeds with ability should select studies which would "equip them to fill a position ... in case the crisis becomes so acute that the national interest demands their services."[48]

The U.S. Office of Education, conscious of a growing manpower crisis, had wanted to draw women into war training from the beginning. Experts worried that "on the average women will require more training than men" since they had less technical backgrounds. As one Cornell instructor put it, women were "handicapped to the extent that by tradition their experiences have been womanly.... They have not had the advantage of playing with Erector Sets and tinkering with Model T's. They have the rather tough job of catching up on things mechanical ... in a relatively brief ... time." But there seemed no other way to meet industry's personnel needs.[49]

In 1943, fifteen colleges across the country offered ESMWT courses entitled "Engineering Fundamentals for Women," intended to help women with bachelor's degrees qualify for junior engineer positions with the navy, War Department, or other civil service. Those courses involved 320 hours' worth of work in engineering math, drawing theory and practice, mechanics of materials, surveying, and shop processes. When Westinghouse realized in 1944 that several of its draftsmen were about to be called into the forces, managers immediately hired fifteen female college graduates and put them through a six-month full-time ESMWT course.

The number of women with degrees in math and science was limited, and the WAACs and WAVES also sought to recruit such women. Schools ran newspaper advertisements and drummed up publicity to interest college students in war training. At Penn State, at least sixty-five women (most majoring in liberal arts, education, or home economics) signed up for classes in airplane and ship drafting. The campus paper editorialized, "We think [that free training of] six hours a week with a definite goal ahead, including an enviable salary, would be worth consideration of any coed. . . . [T]here's much to gain and little to lose by enrolling in these defense courses." Indeed, employers competed to attract participants; during one ESMWT course at Illinois Institute of Technology, companies hired all sixty women enrolled before they even finished class.[50]

In yet another government initiative, the Signal Corps trained over two hundred women for radio engineer jobs in the civil service. Women went to Purdue, Missouri, Illinois, Minnesota, Kansas, and other participating schools for a six-month course in radio theory, the physics of sound, electrical lab work, shop practice, math, drafting, and engineering materials. The women were paid standard wages for forty hours of classwork per week, plus overtime for Saturday study. The "under-engineer trainees," as the women were known, were then assigned to the Aircraft Radio Laboratory at Wright Field in Dayton, Ohio, center of the latest secret radio and radar development. Federal education experts took pride in their success at cultivating women for wartime technical work. One reported, "From talks I have had with young women who are in . . . or finished training, it occurs to me that, although in many cases they are slightly overwhelmed with their first view of the engineering field, they feel that they are getting a much firmer grip upon this man's world into which they are being forced."

War even fostered appreciation of the potential talent of women in home economics, especially those majoring in domestic technology. For years, Iowa State had required household equipment students to take math, physics, and electrical work. In 1942, after Naval Research Lab recruiters came to interview those majors, the home economics department added a five-hour calculus course to help its women enter engineering. At industry's recommendation, the college also organized a special electrical engineering class to prepare home equipment majors for defense employment. Students who signed up were nicknamed WIRES, "Women Interested in Real Electrical Subjects." Professors originally planned to give "these girls . . . elementary background [as] a gentle transition from biscuit baking," but as one instructor wrote, anyone "who expect[ed] to see the girls changing a fuse or repairing a toaster cord [ended up] sadly disappointed. Baby stuff! They learned those things in their own equipment lab when they were freshmen." WIRES were ready for "more rugged topics" such as magnetic circuits, vector diagrams, transformers, and synchronous motors. Though the program turned out only a handful of graduates, those women immediately entered into wartime testing and design work for Western Electric, General Motors, and General Electric.[51]

Desperate for skilled personnel, a number of companies established classes specifically to steer female students into their employ. At the request of Grumman Aircraft, ESMWT instituted an engineering aides' training course at Columbia University in 1942. That course was given five times, and 251 women completed it. The company's recruiting booklet read:

You probably never thought of yourself in engineering—that has long been considered a man's forte. Only a few years ago, we too considered it so. But every day the improbable becomes possible, and we have discovered that a girl with certain qualifications ... can be trained in a relatively short time.... With a dearth of available engineers, we are looking for young women to assist the men we have.... Their satisfactory handling of sub-professional assignments [leaves] our graduate engineers free to concentrate on more complex problems of aircraft design and production.

Grumman accepted women with college degrees in any major (though it preferred math, science, architecture or business) and paid each thirty dollars a week during training. Once women finished at Columbia and entered Grumman shops, they spent three afternoons per week getting additional training in aircraft structural layout.

Other airplane companies soon arranged similar programs. At Penn State, the Hamilton Standard Propellers Division of United Aircraft set up both six-month and year-long courses, training over 130 women as engineering aides. The "Hamilton Propeller Girls," as they were known, were enrolled in Penn State's engineering school, where they studied engineering design, aerodynamics, and metallurgy. At Rutgers, the Eastern Aircraft Division of General Motors ran three sessions of a three-month course in mechanical engineering, metallurgy, math, and shop. Eastern trained about fifty young women, one of whom described the opportunity as "one in a million."

Given the success of training women for aircraft engineering, other industries got into the act. Through Purdue, the Radio Corporation of America ran two eleven-month programs training over 140 women as engineering aides in radio design and quality control. Those trainees worked in Purdue's machine shops, observed RCA's Indianapolis plant, and studied engineering math, materials, electronics, radio circuits, shop practice, and drafting. One faculty member admitted that teaching RCA Cadettes changed his mind; while "three years ago [he] wouldn't have thought so," he came to believe that women definitely had a place in engineering.[52]

General Electric, in an advertisement headlined "Girls, Girls, Girls," announced that it was "hiring young college women to do work formerly done by male engineers ... [to] make computations, chart graphs, and calibrate fine instruments for use in the machine-tool industry.... Although no one expects these girls to become full-fledged engineers, most of them will be given the Company's famous 'test' course." GE recruited women with degrees in math or physics for its on-the-job training, but that pool was limited. Firms began reaching down further, to draw women still in college.

That was the aim of the Curtiss-Wright company, whose planes were a mainstay of the war and which was having trouble meeting production targets. In 1942, the firm developed a plan for training female engineering aides, whom it called

"Curtiss-Wright Cadettes." The company chose seven colleges—Cornell, Penn State, Purdue, Minnesota, Texas, Iowa State, and Rensselaer Polytechnic Institute—to teach a special technical curriculum to a total of more than seven hundred women. Program representatives advertised in college papers and visited schools across the East and Midwest to recruit sophomore, junior, and senior coeds. Curtiss-Wright offered "a unique opportunity to participate in the war effort"; Cadettes would receive $10 per week during their tuition-free, intensive training. Once they were assigned to Curtiss-Wright research, testing, drafting, and production divisions, the women could earn about $140 a month.

A few women selected already had some experience with technical studies (for instance, studying architecture), but most, liberal arts or home economics majors, started cold. Cadettes underwent a ten-month immersion in aviation technology and science. Their course in flight theory taught fundamental aerodynamics, while the strongest students learned engineering math up through calculus. In drafting class, Cadettes practiced detailing actual airplane parts, following Curtiss-Wright standards and typical company blueprints. Studying engineering mechanics and materials, Cadettes used the same textbook as regular students, except that their problems in statics, dynamics, and structural analysis emphasized practical knowledge of airplane construction. To familiarize the women with company procedures, Curtiss-Wright created a class in job terminology and production methods. Entering school machine shops, Cadettes learned welding, soldering, and machine-tool operation to prepare them for shopfloor liaison work.[53]

At participating land-grant campuses, the Cadette program forced faculty to adjust to the sudden arrival of significant numbers of female engineers. Minnesota Cadettes remembered a "reputedly tough professor who strode into his first class and suddenly burst into uncontrollable laughter, eventually recovering to admit that he had never before faced 25 females wielding slide rules." But Cadettes could claim to be doing their part for the war effort, and on those patriotic terms, they were welcomed. Moreover, some skeptics ended up pleasantly surprised by the women's ability. At Penn State, roughly one-third of Cadettes received grades high enough to qualify for the dean's list. Assistant dean G. M. Gerhardt said such performances proved "that these girls could absorb and apply much more engineering training than anyone had anticipated." Instructors reported that the challenge of teaching women with little previous technical experience improved their classroom technique. One said, "I discovered that many things were not instinctively obvious which I had previously taken for granted. . . . Now I throw emphasis on really basic and difficult points. . . . My stock of practical examples . . . is appreciably increased."[54]

Cadettes' presence also led male engineering students to reevaluate assumptions linking technical education to masculinity. Purdue's 1943 yearbook noted, "Tradition . . . seems destined to vanish as the demand for man power opens careers for women in . . . fields heretofore . . . practically uninvaded by the fair sex."

The *Iowa Engineer* went further, editorializing, "Men . . . were on guard for the preservation of the good name of Iowa State's engineering school when Curtiss-Wright . . . brought a group of girls here to study aeronautical engineering. . . . Girls in the wind tunnel lab, in the shop . . . caused the engineers to wonder, then acknowledge, and finally resign themselves to the fact that there would be similar incursions as long as the war continues, and perhaps even after the war."[55]

Schools benefited by participating in the program; when civilian enrollment fell, engineering departments could justify keeping men on staff by assigning them to teach Cadettes. Moreover, the Cadette program made great public relations in a nation pumped up over the war effort. Cadettes proved temptingly photogenic; local papers ran dozens of stories, and *Life* published a special feature. At the end of their crash course, Cadettes received certificates of acknowledgement, then went into Curtiss-Wright plants. Wartime was chaotic; some Cadettes stayed in their initial assignments for only a relatively short time before leaving for other employment or following husbands to a different location. Others remained at Curtiss-Wright for the duration. Though often underused, many appreciated the feeling of contributing to national defense and the relatively decent pay.

Though the Cadette program was a temporary wartime expedient, it helped break down barriers to women's participation in campus engineering culture. At Penn State, Curtiss-Wright delegates served on the Engineering Student Council, the first time women ever participated there. Since they were enrolled in aeronautical engineering, Iowa State Cadettes were eligible to join the campus chapter of the Institute of Aeronautical Science. Purdue Cadettes and Signal Corps women attended local meetings of the American Institute of Electrical Engineers, which "seemed glad to welcome girls into their organization, perhaps for reasons other than merely insuring a big membership." Cadette Marjorie Allen reported that over cocoa and doughnuts, AIEE men got "busy establishing themselves in the good graces of certain Cadettes." In any case, it marked a milestone, the first time that college meetings of professional engineering societies included a sizable representation of women.[56]

More than that, the Curtiss-Wright program marked the first time that enough women were studying engineering to form their own organizations. Students organized a Cadette Engineering Society on each of the seven campuses, with regular meetings featuring movies about aviation and guest speakers on engineering. Members also built model planes, practiced tearing down and rebuilding airplane engines, and discussed topics such as high-altitude flying.

Still more important, the presence of Cadettes and other women in special wartime training reflected a positive light on the growing number of women studying engineering as a regular college degree. By August 1944, Purdue had more than thirty women enrolled in engineering; by December 1945, about seventy-five. A critical mass made life easier; aeronautics major Helen Hoskinson remarked, "Now that lady engineers are not a novelty on this campus, people no longer stare at the sight of a girl clutching a slide rule." More than that, rising numbers helped validate the notion that women could handle technical subjects. Maxine Baker wrote, "We are not asking the men to . . . give up their places to us. We want only to be accepted as co-workers. . . . Let the feminine voice speak loudly."[57]

It was no coincidence that World War II brought a number of "firsts" for co-eds in engineering. In 1944, the *Iowa Engineer* reported, "A woman invaded the Guard of Saint Patrick for the first time in the history of Iowa State College." Civil engineer Ruth Best joined thirty-three men initiated into the scholarly honor society that year. Her selection opened the gate; over the next few years, other female engineering majors also earned membership. Four months later, Eloise Heckert became the first Iowa State woman initiated into Pi Tau Sigma, the honorary mechanical engineering fraternity. In 1945, architectural engineering junior Mary Krumholtz became the first woman to edit Iowa State's engineering magazine. She immediately wrote an editorial saying:

> [S]lide-rule-pushing girls are no longer a rarity. . . . We see them on our own campus and they are not the problem they were once expected to be. In fact, they are a problem only inasmuch as their fellow students and instructors choose to make them one. . . . Obviously there is a long struggle ahead for any woman who presumes to enter a 'man's field.' Men cannot be expected to share the profession voluntarily, and in the dim, distant future when the break does come—and women are accepted rather than tolerated—the concession will be made only as a matter of necessity. Meanwhile we shall continue with our present compromise.[58]

Compromise was a good word for it; growing numbers of female students had not turned engineering departments into a feminist paradise. Some ridicule continued; the 1944 *Cornell Engineer* ran a headline, "WOES (Women of Engineering Schools) Are Here." The piece said, "Rumor has it that 17 woman engineers are at Cornell. Do they build up morale or do they provide distraction? Are they taking advantage of the boy-girl ratio in engineering, are they just trying to help the war effort, or do they want engineering careers?"[59]

Toward war's end, Cornell started to worry about women taking up too much room on campus, as returning veterans tightened up housing. Deans agreed that for the late spring term of 1945, departments should admit no new female undergrads, except in home economics. Cornell's engineering dean had already approved admission of nineteen women; combined with the eighteen women already enrolled, the engineering college thus exceeded its quota of twenty-five women by fifty percent. Cornell's vice president scolded the engineering school for carelessness and stated that absolutely no more female students would be admitted to engineering that semester.[60]

Through the late 1940s and 1950s, students and faculty at schools across the country publicly debated women's place in engineering. In 1955, Penn State engineering dean Eric Walker wrote an article titled "Women Are NOT For Engineering." Walker declared that despite the success of "unusual women" such as Lillian Gilbreth and Edith Clarke, most women did not have the "basic capabilities" needed to handle technical work. He concluded that teaching coeds engineering was not a sound investment, since "[t]he most evident ambition of many women is to get married and raise a family.... Few companies are willing to risk $10,000 on a beautiful blonde engineer, no matter how good she may be at mathematics." Two female engineering students at Florida State jumped to defend their sex, insisting that women's technical skills and professional commitment be respected. In a response headlined "Women Are for Engineering," Wilma Smith pointed out that increasing numbers of women wanted to continue careers after marriage. Penelope Hester added, "If someone can do a job well, why should he or she be denied the right to do that job? An all-male concept of engineering is based on prejudices and old-fashioned ideas.... A woman can be just as devoted to her job as a man, and maybe even more so."[61]

After the war, female engineering students had some momentum behind their claim for recognition. In 1946, about twenty female engineering students at Iowa State organized a local group called "Society of Women Engineers" (four years before the national group of the same name) to assist "in orienting new women students in the division." That same year, female students at Syracuse and Cornell vented their frustration at being excluded from several major engineering honor societies (or restricted to a "woman's badge" instead of full membership). The new honorary society they created, Pi Omicron, soon had chapters at schools around the country. Members held orientations for new female engineering majors and hosted speakers such as Lillian Gilbreth. The mission was "to encourage and reward scholarship and accomplishment ... among the women students of engineering ... ; to promote the advancement and spread of education in ... engineering among women." Then in 1950, female engineers on the East Coast began getting together, officially incorporating in 1952 as the Society of Women Engineers. Significantly, many of the group's early leaders had received their engineering degrees either just before or during World War II.[62]

Though women's foothold in engineering departments remained tenuous, World War II programs made a long-term difference. In subsequent years, the Cold War provided further justification for encouraging young women to pursue engineering. Educators, politicians, and government experts warned that in the atomic age, it would be fatal if the U.S. kept wasting half its brainpower. In 1952, Arthur Flemming, manpower chief in the Office of Defense Mobilization, wrote, "[W]e haven't got a chance in the world of taking care of that deficit of engineers ... unless we get women headed in the direction of engineering schools." Flemming warned, "Soviet Russia isn't making this mistake" of ignoring women's talent.

All told, World War II courses for Curtiss-Wright and other aircraft companies trained about 1,670 women as engineering aides. Hundreds more participated in the RCA, GE, and ESMWT programs. True, most of those who completed wartime training did not make lifetime careers out of engineering, but their experience left its mark on American colleges. A Penn State professor later remembered, "We had [two or three] girls in electrical engineering from the time I got here [in 1931] and I guess they had them before.... But to have groups of them like that!" That was the key difference. Before World War II, the one or two women who occasionally chose to pursue engineering at land-grant schools like Penn State were an anomaly, a curiosity. Wartime programs sponsored by government and private companies suddenly brought their ranks up to a critical mass. With several dozen or even a hundred at a time studying technical subjects, the women could provide each other with crucial intellectual and psychological support.

More than that, World War II gave female engineering students at places like Iowa State, Penn State, and Cornell a collective identity and a chance to build up the numbers over succeeding years. In 1949, there were 763 female students enrolled in engineering at schools across the U.S.; by 1957, that total had more than doubled to peak at 1,783. True, given that the number of male engineering students also soared during those years, women remained less than one percent of total engineering enrollment. But at individual institutions, the difference was apparent. As early as 1946, Cornell had 34 women enrolled in engineering, whereas in previous decades there had generally been no more than about four—and in many years, none. True, for years to come, the campus climate for engineering women would remain chilly, discouraging some to the point of dropping out. But those who had earned their degrees during the 1940s insisted that female students in wartime programs had proven that they could survive in modern engineering education.

Notes

The author would like to thank Iowa State University, the National Science Foundation, and the National Academy of Education for support during preparation of this essay.

1. "Engineers' Day Queen, St Pat Parade Today," *Minnesota Daily*, May 8, 1942: 2.

2. A. H. White, president SPEE to A.A. Potter, U.S. Office of Education, Dec. 20, 1941; Cornell University archives, 16/2/2077, box 7.

3. Raymond F. Howes, "Engineers Shift to War Production," *Cornell Alumni News*, April 9, 1942; 325.

4. Brochure "Essential Facts about ASTP," [ca. late 1943]; Univ. of Minn. archives, AB1.1, Presidential Papers Supplement, box 6, file "ASTP Bulletins."

5. "U.S. Naval Training School at Cornell," *Cornell Alumni News*, August 1942: 469; and "Navy Expands Course," *Cornell Alumni News*, Sept. 15, 1943: 98.

6. "Military Enrollment Increase," *Iowa State Daily Student*, Sept. 30, 1943: 1; and "Military Campus," *The Iowa Engineer*, November 1943: 72.

7. "Recognize Technical Schools," *Iowa State Daily Student*, May 12, 1942: 3.

8. Fred Kelly to A.A. Potter, July 8, 1940; National Archives, RG12 Office of Education, entry 214 ESMWT, box 2, folder "Potter, Dean AA"; and Potter to J.W. Studebaker, July 29, 1940; Nat'l Archives, RG12, entry 222, box 1, folder "vol. 2."

9. J. W. Studebaker, "Proposals to Expand the Program of Training for National Defense Through Schools and Colleges," July 27, 1940; National Archives, RG12, entry 222, box 1, folder "vol. 2."

10. "Tentative Engineering School Intensive Course for the National Defense Program," August 15, 1940; Nat'l Archives, RG12, entry 222, box 1, folder "vol. 2."

11. H. M. Crothers, "Responsibilities and Problems of the Washington Staff," June 21, 1943; National Archives, RG12, entry 214, box 1, folder "Members- Bishop, F.L.—advisory committee."

12. J. W. Studebaker, "Proposals to Expand the Program of Training For National Defense Through Schools and Colleges," July 27, 1940; National Archives, RG12, entry 222, box 1, folder "vol. 2."

13. Studebaker to presidents of engineering institutions, Oct. 18, 1940; and R.A. Seaton to EDT regional advisors, May 2, 1941; National Archives, RG12, entry 222, box 1, folder "3 EDT – letters."

14. Minutes, meeting of regional advisors, Feb. 7, 1941; National Archives, RG12, entry 214, box 2, folder "Potter, Dean A.A."; Vera Christi, "The Job Situation," *California Monthly*, March 1941; and minutes of EDT regional advisors meeting, Oct. 31, 1940; Nat'l Archives, RG12, entry 222, box 1, folder "3 EDT-misc."

15. Minutes of EDT regional advisors meeting, Dec. 21, 1940; Nat'l Archives, RG12, entry 222, box 1, folder "3 EDT-misc."; Memo from R.A. Seaton to institutional representatives of engineering schools, Apr. 15, 1941; Nat'l Archives, RG12, entry 214, box 1, folder "correspondence EDT and ESMDT—Advisory Committee."

16. Prentice to Hotchkiss, no date [ca. Feb. 1941]; RPI Archives, Hotchkiss Papers AC1, box 9, folder 105; Hammond to Seaton, Jan. 21, 1941; Nat'l Archives, RG12, entry 214, box 2, folder "Members—Hammond, H.P."; Minutes, meeting of EDT advisors, Feb. 8, 1941; Nat'l Archives, RG12, entry 214, box 2, folder "Potter, Dean A.A."; Vera Christi, "The Job Situation," *California Monthly*, March 1941.

17. "Defense Program Moves Rapidly," *Penn State Alumni News*, Mar. 1941: 2; and "The Pennsylvania State College Trains for War" pamphlet, no date [ca. 1944]; Penn State archives, GVF/Events/War/WWII/PSC ESMWT Program.

18. D. V. Terrell to Seaton, Mar. 31, 1941; National Archives, RG12, entry 222, box 1, folder "Results of training."

19. R. Randall Irwin to Seaton, Apr. 2, 1941, and L.A. Dane to Seaton, no date [ca. Mar.–Apr. 1941]; National Archives, RG12, entry 222, box 2, folder "Results of training—industry."

20. A. J. Thieblot to Seaton, Mar. 29, 1941, and Joseph Lambie to Seaton, Mar. 31, 1941; National Archives, RG12, entry 222, box 1, folder "Results of training."

21. Anonymous students to EDT offices, no date [ca. Mar.–Apr. 1941]; National Archives, RG12, entry 222, box 2, folder "Results of training—students."

22. Central Iron and Steel Company to Seaton, Mar. 18, 1941; National Archives, RG12, entry 222, box 2, folder "Results of training—industry"; George Wold to ESMDT, Oct. 10, 1941 and Anthony M. Mitchell to ESMDT, Oct. 13, 1941; Nat'l Archives, RG12, entry 222, box 12, folder "vol. 27."

23. Russel McBride to Potter, March 12, 1941; National Archives, RG12, entry 214, box 2, folder "Potter, Dean A.A."

24. Minutes, meeting of EDT advisors, Feb. 8, 1941; Nat'l Archives, RG12, entry 214, box 2, folder "Potter, Dean A.A."; "Resolutions Adopted, Joint Meeting of Regional Advisors and Advisory Comm.," Feb. 8, 1941; RPI Archives, AC1, box 9, folder 105.

25. U.S. Office of Education, "Answers to Questions Pertaining to ESMWT," Sept. 1943; RPI Archives, Houston papers, box 62, folder 645; and minutes, meeting of regional advisors, June 27, 1941; Nat'l Archives, RG12, entry 222, box 1, folder "vol. 3."

26. "Report on Meeting with Commissioner of Education Regarding ESMWT," Nov. 22, 1943; National Archives, RG12, entry 214, box 2, folder "Potter, Dean A.A."

27. E. B. Norris to Seaton, Apr. 7, 1941; National Archives, RG12, entry 222, box 2, folder "Results of training—colleges."

28. Saville to Potter, May 12, 1941; National Archives, RG12, entry 214, box 2, folder "Members- Saville, Thorndike—advisory committee."

29. Memo from the SPEE, Dec. 17, 1941; National Archives, RG12, entry 214, box 2, folder "Potter, Dean A.A."; Minutes, national advisory committee meeting, Dec. 19–20, 1941; RPI Archives, Hotchkiss Papers AC1, box 9, folder 105.

30. "Science Dramatized for Wartime Students," *Penn State Alumni News*, July 1942: 10.

31. Penn State ESMWT booklet and "The Pennsylvania State College Trains for War" pamphlet, no date [ca. 1944]; Penn State archives, GVF/Events/War/WWII/PSC ESMWT.

32. Minutes, meeting Texas Institutions with Regional Advisor, March 5, 1942; Univ. of Texas archives, box 4R418, folder "War training records, ESMWT, Texas committee minutes 1941–45"; "Train Ordnance Workers" *Cornell Alumni News*, Mar. 5, 1942: 264.

33. Potter to H.H. Armsby, Sept. 8, 1942; Nat'l Archives, RG12, entry 214, box 2, folder "Potter, Dean A.A."; R.A. Seaton to ESMDT institutional representatives, Oct. 8, 1941 and George Case to ESMWT representatives, Jan. 19, 1943,

AC1, box 11, folder 139, RPI archives; "Students Learn Radar Secrets," *Iowa State Daily Student*, July 30, 1943: 1; "RADAR Training Given," *Iowa Engineer*, Mar. 1943: 166.

34. O.W. Towner to ESMDT, Oct. 16, 1941; National Archives, RG12, entry 222, box 2, folder "letters vol. 3"; and Minutes, national and regional advisory committee meeting with industry representatives, June 29, 1942; National Archives, RG12, entry 214, box 1, folder "Meetings—industry, guests from—Advisory Committee."

35. E. R. Torgler to ESMDT, Nov. 1, 1941; National Archives, RG12, entry 222, box 2, folder "letters, vol. 3"; and Notes, *Penn State Alumni News*, June 1941: 21.

36. William Urschel to Seaton, Apr. 4, 1941; National Archives, RG12, entry 222, box 2, folder "Results of Training—Industry."

37. University of California to ESMWT, no date [ca. Aug. 1942]; National Archives, RG12, entry 222, box 13, folder "Replies to misc. 45"; and ESMWT newsletter, Sept. 30, 1942; National Archives, RG12, entry 214, box 2, folder "Potter, Dean A.A."

38. Clarence E. Heitman to ESMDT, Oct. 18, 1941; National Archives, RG12, entry 222, box 12, folder "vol. 27." The University of California also ran electrical courses inside the Permanente Metals shipyards, a supervisors' training course at Moore Dry Dock Company yards, two classes in plastics at the Alameda Naval Air Station, and a foremen's training course at the Pacific Gas and Electric Company. ESMWT Newsletter, July 31, 1943; RPI archives, Houston collection, box 62, folder 645.

39. Ira Abbots to ESMWT, Jan. 3, 1942; National Archives, RG12, entry 222, box 12, folder "letters vol. 4."

40. E. B. Norris to Seaton, Apr. 7, 1941; National Archives, RG12, entry 222, box 2, folder "Results of training—colleges"; F.M. Tompkins to Seaton, Mar. 25, 1941, and M.P. Auburn to Seaton, Mar. 27, 1941; National Archives, RG12, entry 222, box 1, folder "Results of training."

41. Roger M. Blough, "The Challenge of America," Jan. 30, 1956; Cornell Univ. archives, Hollister papers 16/2/2077, folder 43-25.

42. Amy Bix, "Feminism Where Men Predominate: The History of Women's Science and Engineering Education at MIT," *Women's Studies Quarterly* 28, no. 1 & 2, (Spring/Summer, 2000): 24–45.

43. Raymond Howes, "Concerning Sibley Sues," *Cornell Alumni News*, Oct. 13, 1938: 30.

44. "Beauty Meets Resistance," *The Penn State Engineer*, October 1934: 9; and "Honors Day," *The Iowa Engineer*, November 1935: 37.

45. Charlotte Bennett, "An Opinion on Engineering for Coeds," *Purdue Engineer*, Nov. 1935: 33.

46. Ellen Ziegler, "Women in War Industry," *Purdue Engineer*, July 1942: 26.

47. "Guidance for Girls," *Purdue Exponent*, July 31, 1942: 2; and Genevieve Husted, "AWS Will Hold First Vocational Meeting for Coed Scientists," *Purdue Exponent*, Oct. 15, 1942: 1.

48. "University to Test Science Aptitudes of Coed Students," *Purdue Exponent*, Dec. 1, 1942: 1; "Coed Conclusions," *Purdue Exponent*, Dec. 8, 1942: 2; "Hockema Expresses Desire for Coeds to Remain Here," *Purdue Exponent*, Dec. 10, 1942: 1.

49. Fred Ocvirk, "Cadettes at Cornell," *The Cornell Engineer*, April 1943: 8–9; and Minutes, national advisory committee meeting, June 29, 1942; National Archives, RG12, entry 214, box 2, folder "Potter, Dean A.A."

50. "We the Women," *The Daily Collegian*, Oct. 10, 1942: 3.

51. "Interview Women for Naval Jobs," *Iowa State Daily Student*, Oct. 9, 1942: 3; "Require Equipment Majors to Add 5-Hour Course," *Iowa State Daily Student*, Oct. 15, 1942: 4; Ben S. Willis, "The Wires Take Over," *Iowa Engineer*, Oct. 1943: 41.

52. "George Cooper," *Purdue Engineer*, April 1946: 38.

53. *War Training programs—WWII—Curtiss-Wright Engineering Cadette Training Program*, (Ames, IA: Iowa State College, 1945).

54. *Fifty Years of Aeronautical Engineering: University of Minnesota, 1929 to 1979*; "Lady Engineers Upset Tradition," *Pittsburgh Press*, May 17, 1944: 5; and Warren Bruner, "A Report on the Engineering Cadette Training Program of the Curtiss-Wright Corporation," RPI archives.

55. *The Purdue Debris*, 1943: 20.

56. Marjorie Allen, "Cadette Column," *Iowa State Daily Student*, March 5, 1943: 2; "Girls Invade AIEE Meeting, Hear Hale Speak on Airfields," *Purdue Exponent*, May 29, 1943: 5; and R.M. Harnett to Anne Blitz, Oct. 15, 1942; Univ. of Minnesota archives, AW9.1, Dean of Women's Papers; folder 27, "Women in War."

57. Peggy Stefen, "Coed Engineer Comes Through Over Comments of ME Faculty," *Purdue Exponent*, July 22, 1943: 1; and Maxine Baker, "What About the Women?" *Purdue Engineer*, December 1945: 5.

58. Mary Krumholtz, "Women in Engineering," *Iowa Engineer*, May 1945: 176. "Guard Pledges Woman," *Iowa Engineer*, Jan. 1944: 100; "Pi Tau Sigma Elects," *Iowa Engineer*, April 1944: 155.

59. "WOES (Women Of Engineering Schools) Are Here," *Cornell Engineer*, Jan. 1944: 13.

60. George Sabine to S.C. Hollister, Nov. 22, 1944 and July 4, 1945; Hollister to Sabine, July 2, 1945; Cornell Univ. archives 16/2/2077, box 44, 71.

61. Eric Walter, "Women Are NOT for Engineering," *The Penn State Engineer*, May 1955: 9, 20; Wilma Smith, "Women Are for Engineering," and Penelope Hester, "What Really Counts?" *The Penn State Engineer*, April 1956: 18–19.

62. "New Society Organizes," *The Iowa Engineer*, May 1946: 222.

3 Introspection: Land-Grant Engineering's Past as Future

Reengineering the Land-Grant University

The Kellogg Commission in Historical Context

Howard P. Segal

In American education, as in other realms, cries of "crisis" invariably garner more public attention than calmer pleas for public response to alleged problems. In education, moreover, it has become a veritable tradition in rhetoric, if not in reality, that every generation of students, teachers, and administrators faces ever graver challenges than all prior generations. This has long been not just the stuff of commencement addresses but also the thrust of more sustained arguments. Of course, not everyone agrees with these analyses. When, for example, in 1963 then President Nathan Pusey of Harvard published a collection of essays about the alleged awful crises facing his illustrious and wealthy institution—above all, declining student attendance at Memorial Church in Harvard Yard—most of the academic world did not see things quite so gloomily as Pusey did.[1] Published before the assassination of President Kennedy later that year, and before the student upheavals at Harvard and elsewhere later in the decade, the book was quickly forgotten. There are countless other such works that may today be found in academic libraries and used bookstores.

But the sense of crisis certainly has a higher ring of truth when the rest of the "real" world also seems to be changing, for better or for worse. In our day, when, we are repeatedly told, technology is transforming everything, it is no accident that leaders of higher education should have joined the ongoing debates over future trends. Consequently, there have already been and will continue to be endless speeches, conferences, and publications about high-tech and higher education.

The most significant and most systematic response to date to the changing landscape of higher education has been the Kellogg Commission on the Future of State and Land-Grant Universities. The Commission was established in January 1996 by one of the nation's leading higher education organizations: the National Association of State Universities and Land-Grant Colleges (NASULGC). In an earlier embodiment, NASULGC itself dates back to 1887 and is the country's old-est higher education organization. The Commission's funding came from the highly respected Kellogg Foundation of Battle Creek, Michigan. As the Commis-sion acknowledged at the outset, its creation was due in part to the end of the Cold War and the need and opportunity to reallocate national resources to do-mestic problems (this was before the resurgence of concern over supposedly di-minished Pentagon budgets and over the absence of a supposedly workable "Star Wars" anti-missile defense system). But, more specifically, the Commission was set up in response to a "chronic shortage of funds, public demand for greater ac-countability, soaring fees, and hard questions about research and faculty work-load." The Commission was charged by its sponsors with nothing less than trying to "define the direction public universities should go in the future and to recom-mend an action agenda to speed up the process of change."[2]

Composed of the presidents and chancellors of some twenty-five colleges and universities, the Commission issued six reports, the final one appearing in March 2000, coinciding with its final major meeting. These reports must be taken seriously, not only in and of themselves but also because of what they reveal about American higher education today. And because NASULGC institutions award roughly a half-million degrees annually, including about 33 percent of all United States bachelor's and master's degrees, 60 percent of all doctoral degrees, and 70 percent of all engineering degrees, the Commission's findings and recommenda-tions may have profound effects on American higher education—and American society—overall.[3]

Of particular interest here are, first, the Kellogg Commission's depiction of how contemporary technology is affecting higher education; second, its sense of historical change, especially its depiction of the evolution of land-grant institu-tions and the land-grant ethos from the crucial 1862 Morrill Act until the present; and third, its envisioned role for engineers and engineering in the future. I believe that, in all three cases, the Commission is representative of American higher edu-cation in general, and that the dilemmas it faces—or, more often, ignores—are also representative of higher education in general. I also believe, and will elaborate later, that these three aspects collectively constitute a de facto social engineering project for American higher education and American society that is the Commis-sion's real vision, a vision intended to restore to land-grant and state colleges and universities a crucial role in transforming the nation. Finally, I also believe, and will repeatedly note, that the Commission's persistent demand for reducing faculty from an intellectual elite to the virtual peers of their students never extends to the top ad-ministrators who increasingly constitute the actual academic elite. It is no accident that not a single faculty member (or student) served on the Commission.

Technological determinism—the notion that technology shapes society and culture—pervades the Commission's reports as much as it does contemporary high-tech hype elsewhere. Like high-tech manufacturers, advertisers, and, not least, prophets, the reports see high-tech advances as transforming the nation and, in due course, the entire world to an unprecedented extent and, no less important, at an unprecedented speed. (The fact that no professional historian of technology accepts technological determinism might have given the Commission some pause. But, of course, historians of technology or of anything else were not consulted, even for the periodic references to the history of higher education.) In reading passages of these reports one is repeatedly reminded of the worldviews of such high-tech prophets as Alvin and Heidi Toffler, John Naisbitt and Patricia Aburdene, Michael Dertouzos, Nicholas Negroponte, and Bill Gates. As the Commission sees it, the greatest challenge facing American higher education today is responding to these high-tech advances—or, in the words of a preliminary 1996 Commission document, "Taking Charge of Change." That the Commission was, as noted, simultaneously charged with "speed[ing] up the process of change" is one of several apparent conflicts throughout the Commission's work that was never resolved, perhaps never recognized.

Like those high-tech gurus, so too the Commission members are at once dazzled by contemporary scientific and technological developments and fearful that, as in a kind of updated *Frankenstein* tale, those developments will end American higher education as we have known it in the twentieth century, above all the ethos and achievements of the land-grant movement. (The fact that many institutional members of NASULGC and in turn the Commission are not official land-grant schools is not a problem insofar as the issues transcend land-grants themselves yet are rooted in the land-grant ethos.) As the initial chair of the Commission put it, "We will either be the architects of change or we will be its victims. There has been a major seismic quiver out there, and we in leadership positions must become tuned to it or become increasingly marginalized."[4]

The Commission is quite clear about that ethos and those achievements. As "Taking Charge of Change" contends, "The nation's state and land-grant colleges and universities have promised many things to many people and delivered on most of them: world-class research, first-rate service, and access to affordable education for all. They have been a unique source of practical education and lifelong learning, first for farmers and then for just about everyone else. The value of these institutions is beyond calculation. They have kept the promise."[5] This is about as much historical context as one finds in any of the Commission reports and other documents, beyond the repeated references to the Morrill Act of 1862. The sole exception is the call for a "Higher Education Millennial Partnership Act" for the twenty-first century that would complement the Morrill Act and other legislation crucial to the evolution and well-being of public higher education.[6] One would never guess from this and related Commission statements that the land-grant movement has been the subject of intense scholarly as well as political discussion since 1862; that the principal motivations for the Morrill Act have themselves been hotly debated by historians and others for decades; and that this retrospective consensus on promises made and kept is thus in some measure an invention.[7]

Ironically, in terms of the actual historical record, Allan Nevins's 1962 pamphlet *The Origins of the Land-Grant Colleges and State Universities* provides a sense of events as almost overwhelming the framers of the Morrill Act that parallels the similar sense found in the Kellogg Commission:

> It was remarkable as a profession of faith in the future in the midst of civil war; but it was still more memorable as an embodiment of the whole democratic dream of the time and the conviction that the nation must move fast to avoid a betrayal of its imminent needs. Ever in the minds of the leaders was a vision of the families of bright children, springing up by the million over prairie, plain, and foothill, hungry with a New World appetite for knowledge, wisdom, and inspiration. They could no longer be properly served by the small endowed colleges that besprinkled the land—and not by them alone. . . . They needed a new education for a new society. . . . Men should not lose a day in striving to create this new education. Should they stop to draw up careful blueprints?—they had no time. They could plan it as it grew. Would they not make costly mistakes?—certainly; as always; but they would learn while they blundered, and they would meanwhile train a new generation in at least semisatisfactory fashion for the paths it should tread.[8]

To be sure, the Kellogg Commission is hardly characterized by a comparable lack of blueprints. But the Commission's sense of transforming change throughout the nation and its sense of urgency in needing to respond have parallels to Nevins's portrait of the situation in 1862. So, too, of course, does the Commission's sense of millions of prospective students needing to be attended to, as discussed below.

The Commission's actual ambivalence about technological and other contemporary developments is revealed in "Taking Charge of Change":

> We cannot sugar-coat the truth. Unprecedented problems confront our campuses. We face seismic shifts in public attitudes. We are challenged by new demographics and exploding technologies. We are beset by demands to act "accountably" toward students, parents, communities, and taxpayers. An increasingly skeptical press questions our priorities. . . . Institutions ignore a changing environment at their peril. Like dinosaurs, they risk becoming exhibits in a kind of cultural Jurassic Park: places of great interest and curiosity, increasingly irrelevant in a world that has passed them by.[9]

Yet higher education, that same document continues, has always been "an agent of change. It has been incessantly dynamic and responsive, in both form and substance." In fact—and somewhat at odds with the alleged uniqueness of the contemporary crisis—is the line that "The issue has always been the same—not should we change, but how?"[10] In this context, one might just as easily cite a sentence in Edward Eddy's classic 1957 book on the land-grant idea, *Colleges for Our Land and Time:* the United States "demands of its public colleges the ability and the desire to adapt themselves continually to changing times and, where possible, to anticipate society's needs."[11]

Should one therefore be optimistic or pessimistic? Or somewhere in-between? No less important, why not invoke genuine historical context and suggest that the problems facing the land-grant institutions in their early days were no less severe than those they and other public institutions face today? One need not be a professional historian to concede this. Indeed, the title of most of the Commission's reports is "Returning to Our Roots." Yet to do so would, of course, undermine the fundamental position of uniqueness, of claiming that today's challenges and crises are truly unprecedented. In effect, the Commission seeks the best of both worlds: avowed connections to the past that do not bind too tightly so that the present and the future can somehow simultaneously reduce the past to simpler, less challenging times.

One reason behind this position is surely political: the greater the alleged crisis, the greater the prospects for not just sympathy but outright financial support from government and, increasingly, the private sector. But there is, I suspect, another, no less genuine reason: ideological. The Kellogg Commission was ultimately convinced that, for all their risks to traditional higher education, high-tech advances are not simply inevitable but generally beneficial; that they will allow for the reengineering of land-grant and other public colleges and universities in ways that serve top administrators as well as business/corporate America, local and state, if not national governments, K–12 school systems, engineers and other technical professionals, and, not least, self-conscious student consumers.

Ironically, it may well be the "exploding technologies" celebrated in some of the Commission reports—and barely questioned, but never criticized, in others—that pose the gravest threats to the land-grant institutions in their traditional structures: threats that could topple, not just transform, those and many other public colleges and universities. In particular, the growth of distance learning, especially distance learning requiring little if any presence on conventional campuses, and the parallel rise of virtual universities threaten the need for those conventional campuses and for traditional means of "delivering" instruction. The ultimate logic of these high-tech advances (the immediate downsizing and eventual elimination of all but the most elite college and university campuses) can be drawn from a 1998 speech entitled "The Future of Land-Grant Universities" delivered—under none other than NASULGC/Commission sponsorship—by the president of the University of Illinois, James Stukel. Only the latter institutions would confer degrees whose value in the marketplace would justify students coming to and remaining for some years on campus. The vast majority of American colleges and universities would continue, if at all, merely as administrative sites for distance and virtual learning, with a few possible exceptions such as laboratories.

Stukel, let me emphasize, never makes these points. Far from it: himself a Commission member, his tone is upbeat, so long as mainstream higher education learns to master the Internet, under which he includes "telecommunications and computers." If mainstream higher education can do that, a wonderful new world awaits us, from the assembly and reconfiguration of "virtual research teams," with endless partnerships with business and industry, to the merger of "on-line extension information holdings across the country . . . into a national website" that will transform cooperative extension and, in due course, much else. Within a decade, Stukel predicts, higher education's research in general will be conducted almost entirely "in this virtual environment, too. Think of that. When you have a problem, no longer will you phone your local university or your local cooperative extension service." Instead, a $500 desktop computer "will give you access to the most advanced computing and database environment in the world."[12]

If, according to Stukel, every American citizen could have such a computer and such software, "racial, ethnic, and gender conflicts could be at least eased,"[13] because everyone would then have access to higher education—far more than in the late-nineteenth- and early-twentieth-century "golden age" of the land-grant movement. (That "golden age" is reflected in comments like Eddy's in 1957 that the land-grant colleges and universities had by then provided their campuses with "a cross-section of American life" and had "become an academic melting pot of all classes and kinds."[14]) Access, as elaborated upon later, is one of the key issues the Kellogg Commission deals with. But so, too, are the on-campus experiences of students and faculty alike. Yet will public colleges and universities still lure students if so many can so easily earn their degrees off-campus? Similarly, will the "lifelong learning" also advocated by the Commission be accomplished as much on-campus as off? These fundamental issues are hardly addressed.

In a style akin to Alvin Toffler, Stukel invokes the metaphor of the wave of the future in describing high tech: one either rides it out—and, to this extent, controls one's destiny—or drowns. There is no middle ground. As Stukel puts it, "To use a nautical analogy, I see technology and change as a tidal wave. It is only a very small wave right now," though the rest of his speech would suggest a much bigger wave already. "What we have to do is to turn the bow of our boat into that approaching wave and not be broadside to it and rolled over by it."[15] The ambivalence about contemporary technological and other developments that characterizes some Commission materials is, in effect, repressed, replaced by a reassurance that higher education will not merely survive but flourish. Any opportunity for providing a more mature middle ground is thereby lost. Instead, one finds here, as in so much of the Commission reports and documents, a reassurance reminiscent of the American style of "Positive Thinking," associated with, most famously, the late minister Norman Vincent Peale.

Nevertheless, it remains puzzling that those like Stukel, with a vested interest in keeping traditional campuses flourishing, would ignore what appear to be the obvious consequences of unadulterated distance learning, off-campus cooperative activities and research, and the like. It is one thing for wholly commercial enterprises like the University of Phoenix to encourage this, quite another for university presidents and other educational leaders to do so.

Moreover, while Stukel's suggestion for $500 desktop computers for every citizen is certainly commendable, his—and many others'—questionable subtext is the increasingly common mis-equation of sheer access to information with the acquisition of genuine knowledge. "Information overload" has become part of our language for a very good reason, and the best Web search engines do not eliminate this dilemma. Stukel's proposal, in fact, is reminiscent of those "techno-fixes" that were common in the 1960s and 1970s as government and corporate technocrats struggled to solve the problems of poverty, crime, race riots, and other affronts to affluent Americans by proposing free electricity and free air conditioners for inner-city residents. "Techno-fixes" avowedly substitute for political, economic, psychological, and cultural solutions and are themselves manifestations of that ideological belief in technological progress that, as indicated, pervades the Kellogg Commission.

Not surprisingly, a related subtext of Stukel's—and many others'—visions is that education today is as fundamentally technological a process as it is an intellectual one and that technical experts are therefore at least as important as conventional faculty in curriculum design and implementation. Many online courses at virtual universities like the University of Phoenix are taught by instructors with presumably sufficient technical expertise in the mechanics of online instruction, but not necessarily sufficient academic training—and, in any case, often with little or no control over course content. They might well teach courses developed entirely by others, persons the instructors have never communicated with, much less met.[16] Equally significant, course content no longer seems to matter a great deal, so long as the "delivery system" offers the desired "product" as quickly, as efficiently, and as cheaply as consumers demand. (Designing and operating the "delivery system" and the "product" is one of the opportunities for engineers—and other technical professions—as elaborated below.) Indeed, the Commission repeatedly quotes Justin Morrill's statement that "I would have learning more widely disseminated" to justify such assumptions.[17]

That in turn leads to the Commission's barely qualified de facto endorsement of the consumer culture that, with the encouragement of many college and university presidents and other educational leaders on campus and off, has come to govern almost everything. As the Commission's final report, *Renewing the Covenant: Learning, Discovery, and Engagement in a New Age and Different World*, itself concedes about higher education in America today:

> What is most troubling about all of these developments is that they reveal an apparently growing public consensus that education is simply another commodity, another market for consumers, in which students are customers. Re-

> search, if it is thought of at all, is prized far more for its commercial promise
> than its capacity to extend human capabilities or push back the boundaries of
> our knowledge and understanding.[18]

Despite this admission, there is precious little in the Commission's analyses and recommendations that would attempt to alter this trend and much more that (no pun intended) buys into it wholesale.

Consequently, traditional faculty are boxed in between high-tech-oriented administrators like Stukel, who generally do no teaching of any kind, and students who demand access to instructors and courses at their utter convenience. The motto of such institutions might well be that of clothes discounter Sy Syms: "An Educated Consumer is Our Best Customer." As a colleague of mine with vast experience in high-tech curriculum and instruction recently suggested, any administrators who claim that "putting courses on the Web is easy, straightforward, and inexpensive" should themselves "be required to put a course online—over a weekend."[19]

The use of business terms like "delivery system" and "product" is, of course, no accident, for higher education in America models itself ever more on corporate America. And the Kellogg Commission is hardly an exception. The steady loss of faculty autonomy in curriculum development that stems from embracing a student-centered consumer culture is matched by an equally intense desire to please business and industry. One wonders if any Commission members are familiar with Thorstein Veblen's *The Higher Learning in America* (1918), the pioneering and, by now, classic critique "of business control in practically every aspect of the modern university."[20] As in Veblen's day, so here, this pandering to business and industry goes far beyond wishing to cultivate the financial and political support of the private sector amid the declining financial and political assistance offered by most states in recent decades. Interestingly, Eddy's 1957 work noted complaints from labor leaders about the lack of land-grant extension programs and research for their workers, as opposed to the situation in agriculture and business.[21] The Commission appears equally indifferent to labor's needs.

In such a climate, many faculty increasingly feel akin to the nineteenth-century skilled workers whom David Montgomery discusses in his *Workers' Control in America* (1979). As Montgomery details, skilled workers in pre-assembly line, pre-automated industries once controlled the pace of work and in turn the output because they understood more about the machinery than their managers and owners and thus could not easily be replaced. As the radical union organizer whom Montgomery quotes put it so well, "The manager's brains are under the workman's cap."[22] Similarly, traditional faculty once controlled curriculum and instruction because they knew more about these than deans, provosts, presidents, and trustees and thus commanded considerable respect and deference. But most skilled workers gradually lost control as ever larger and more powerful machinery determined the pace and output of work, as "deskilling" steadily reduced the value and application of their expertise, and as manual labor became identified with mindless labor. The analogy to higher education today obviously has its limitations, but the growing subservience of traditional intellectual work—both teaching and research—to high-tech mechanized work is, I believe, a striking comparison. No wonder, then, that "productivity"—or "accountability," to use the Commission's term—has become a buzzword for educational leaders as much as for governors, legislators, corporate executives, and the general public. The aptly named publisher Productivity Press, hitherto oriented toward (purely) commercial enterprises, now markets such books as *The Visual Factory, Implementing a Lean Management System*, and *Performance Measurement for World Class Manufacturing* as tools for the Press to become "the link between the academic world and the manufacturing/corporate areas, providing educators and students with powerful and strategic information." Its book catalog is subtitled *Productivity for the Academic World*.[23]

Yet the Commission's emphasis in some of its reports on faculty doing more—and more collaborative—teaching in order to be more productive and more accountable may be at odds with its proclamations elsewhere of the need for greater research, especially research that will be useful for local, state, and national businesses, industries, governments, and K–12 school systems. On the one hand, the Commission contends that traditional teaching styles placing faculty above their students both physically and intellectually are harmful to most students— that is, students demanding customer satisfaction—and ought to be replaced by faculty and students working together on a more level playing field, whether in reconfigured conventional classrooms or, perhaps better, online. "Faculty, in this conception, change from being the source of all knowledge, 'the sage on the stage,' to mentors helping lead students toward new understanding, 'the guide on the side.'"[24] On the other hand, the Commission wants faculty to serve the community and the state by developing ever more partnerships with those areas' businesses, industries, governments, and K–12 school systems. Stukel's $500 desktop computers may well serve both teaching and research, both students and nonstudents, but even the most enthusiastic and dedicated faculty have to make hard choices in allocating their time. Their time is not unlimited, however much they may wish to serve their students and their communities and states. The Commission doesn't appear to grasp that fact any more than it grasps the logic of moving so much instruction and service off-campus despite wanting to retain traditional campuses. The most the Commission says is that faculty today "are trapped in a Catch-22 not entirely of their own making. Criticized for neglecting teaching, professors also have to contend with an institutional focus on research reputation that encourages them to emphasize scholarship and research (and accompanying recognition from professional societies and peers) as the dominant reward-and-recognition structure."[25] This is certainly true, but a $500 desktop computer won't eliminate the dilemma. Once more, those who advocate pulling faculty off their alleged pedestals do not step down from their own, much less themselves seek to engage in *any* teaching, research, or service.

In 1992, John Pesek, a distinguished emeritus professor of agronomy at Iowa State, was quoted at length in a controversial opinion column in *Feedstuffs* entitled "Do Land-Grant Schools Do Too Much Research?" Pesek said definitely yes. According to him, "The problem started in the 1950s, . . . when all land-grant colleges wanted to become universities, which required more money than most colleges had. . . . When the colleges were short on funds, they cut back on their [traditional] duties to teach agriculture, engineering, and home economics, and they [instead] increased their reliance on research to generate the attention and funds they needed to expand to universities." By contrast, the editors hoped that vital research in agribusiness—and, by extension, in engineering and other fields—could somehow still be reconciled with good teaching and that properly educated students could, upon graduation, promptly "move into agribusiness."[26]

Pesek's historical perspective may have been a bit simplistic, but he did at least attempt some historical context, which, once more, the Kellogg Commission largely ignores. As Earle Ross's classic work, *Democracy's College: The Land-Grant Movement in the Formative Stage* (1942), reminds us, the potential tension between teaching and research was well recognized at the outset of the land-grant movement, and the efforts made back then to alleviate it may still be useful today. According to Ross, "It was the view and aim of the pioneers . . . that the teaching and research functions were to be combined." Yet "Distinction was usually made between experiments that would have a direct popular appeal through their immediate results and those involving more intricate and prolonged investigation."[27] Both undertakings were eventually deemed valid.

By contrast, as a representative example, the 1997 inaugural address of the then president of the University of Maine, Peter S. Hoff, simply asks for his institution's future research to focus on "stimulating the economy; promoting the well being of existing business, industry, agriculture, and aquaculture; managing and protecting our environment; reinforcing the quality of K–12 education; and emphasizing work directed at technology transfer that promotes and attracts new economic ventures and creates new jobs." Not only was Hoff a member of the Commission, but his inaugural address was entitled "Back to the Future: The University of Maine and Its Land Grant/Sea Grant Tradition." Yet nowhere in his address is there any recognition of a potential tension between these objectives and those for making teaching ever more student-centered, ever more a case of faculty teaching on demand. It is reassuring to ask, as Hoff does, that all programs be "student-oriented in their outlook and in their operation" and that they "focus on what students need to learn rather than on what we prefer to teach."[28] But who determines what students "need to learn" and what happens if business, industry, government, and K–12 schools in Maine change their agendas? Does the curriculum automatically change in turn? And what if, God forbid, the students as all-powerful consumers wish to retain the existing curriculum and programs despite changes in the "real world" or, equally distressing, wish to alter the existing curriculum and programs in the face of opposition from business, industry, or government, or K–12 schools?

The Kellogg Commission, however, transcends this dilemma, among others, by redefining the mission of public universities from "one of teaching plus research plus service to one of '*integrated* learning, discovery, and engagement.'"[29] As Eddy had conceded in 1957, land-grant institutions had "not always been successful in the integration" of teaching, research, and service. Meanwhile, he went on, not a few educational leaders had periodically concluded that land-grant institutions could not simultaneously be ones "of higher learning and . . . operate effectively in the marketplace."[30] For the Commission, "integrated learning" means acknowledging that "learning is not a spectator sport" and that "students take responsibility for their own learning" while collaborating with faculty and fellow students.[31] It also presumes and promotes "lifelong learning" by every citizen physically and intellectually capable of pursuing it. "Discovery" means the redefinition and expansion of "scholarship" as proposed in Ernest Boyer's *Scholarship Reconsidered* (1990), whereby traditional research—"the scholarship of discovery"—is to be joined by "other forms of scholarship—teaching, integration, and application," all three of these to be put on the same footing as the first.[32] And "engagement" means going well beyond conventional extension and service to establish genuinely two-sided partnerships with individuals, groups, other institutions, and communities for mutual benefit. No longer can we preserve the "one-way process of transferring knowledge and technology from the university (as the source of expertise) to its key constituents."[33]

As land-grant and other public universities undergo various forms of reengineering, these three reconceptualizations are not without value. For example, as the Commission suggests, "integrated learning" can lead to some research collaborations on the part of advanced undergraduates with faculty and graduate students. Similarly, adopting Boyer's broadening of "scholarship" can reduce the pressure on junior faculty at many institutions to do only that research leading to traditional publication and tenure and/or promotion and instead allow them to pursue those "other forms of scholarship" that hitherto have received little if any recognition and rewards. Likewise, two-sided "engagement" with institutions' potential external partners surely enhances appreciation of and support for public universities.

But the question remains: who decides how to reconceptualize and then implement these institutions' traditional missions? It is a measure of how much things have changed since Boyer's work appeared in 1990 that he took for granted that "when all is said and done, faculty themselves must assume primary responsibility for giving scholarship a richer, more vital meaning. Professors are, or should be, keepers of the academic gates. They define the curriculum, set standards for graduation, and determine criteria by which faculty performance will be measured—and rewarded."[34] By contrast, the Commission, despite acknowledging "shared governance" between faculty and administrators as "the distinctive administrative feature of our institutions,"[35] still assumes that Boyer's broadening of "scholarship" *will* be accepted (or imposed). The remaining issues to figure out are faculty incentives and rewards for doing so. If this is the situation with research/scholarship, what role will faculty play in redefining teaching and service, or "integrated learning" and "engagement"? In the case of the latter, the Commission makes clear that "an engaged university cannot be brought into being with a 'service' requirement in tenure and compensation that can be met solely through service on campus committees or disciplinary [i.e., professional] organizations. . . . if engagement means anything at all, it reaches beyond the campus and the disciplines that shape it."[36] There would not appear to be much choice here either on the part of the faculty.

Related to this is the Commission's concern with the fragmentation of the university community and efforts to establish, or reestablish, greater unity among its various "cultures." In the words of its report *Returning to Our Roots: Toward a Coherent Campus Culture:*

> . . . today's university community no longer has a single "culture" but several: an academic culture, made up primarily of faculty and students, fragmented into its own subcultures organized around disciplines, self-governing departments, and professional schools; a distinct and entirely separate student culture, with a bewildering diversity of aims and interests, from fraternities and sororities to student associations and research clubs; an administrative culture that tends to be separated from that of the faculty and sometimes in competition with it; and an athletics culture, perceived to be autonomous and beholden to commercial interests.[37]

No one could seriously deny the accuracy of this description, but surely even the original land-grant institutions faced some degree of fragmentation despite their greater consensus on mission, their smaller size, and their less diversified student, faculty, and administrative ranks than today. And as those institutions grew, and as additional land-grant and other public institutions were established, a greater degree of fragmentation ensued. The Commission's implication here that this contemporary fragmentation is only a recent development is yet another reflection of its lack of sufficient historical context.

Not surprisingly, the "administrative culture" receives the least attention from the Commission. Its phenomenal growth in the past few decades is ignored, as is its emergence as a professional culture in itself, with a sense of superiority to the academic culture from which, as noted, it generally seeks distance, not unity. Far from conceding any administrative lapses in terms of undermining traditional faculty power, the Commission criticizes faculty for "pervasive cynicism on the part of some members of the faculty toward the reform agenda of the 1990s and administrators' motivations in advancing change." Far from acknowledging that huge central administrative bureaucracies—often coming at the expense of every other campus "culture"—may not be in universities' best interests, the Commission insists that "administrative costs" are necessary for "fiscal responsibility."[38] Indeed, by "administrative culture" the Commission emphasizes professional and classified employees rather than deans, vice presidents, provosts, and presidents/chancellors. The latter presumably are once more almost beyond criticism.

Yet it is the latter, of course, who intend to bring these cultures together. True, as the Commission states, "respecting faculty integrity does not require hamstringing central administration." True, too, governance has too often been treated by everyone as a "zero-sum game, in which authority gained by one of the three [governing boards, administration, and faculty] comes at the expense of the other two."[39] But it is curious, to say the least, that the Commission's proposed reengineering of land-grant and other public universities in ever more democratic directions for every other constituency does not appear to extend to those at the top.

Ironically, despite all the high-tech hype in the Kellogg Commission reports and documents, engineering education itself is not much discussed. But then, in fairness, neither are the curricula of any other fields. The notable exception is a call for more interdisciplinary learning, discovery, and engagement.[40] Otherwise, the focus, as discussed shortly, is on the hardware and software necessary to deliver information and instruction anywhere at any time to anyone. Hence, to repeat, the actual intellectual contents of the "product" is at best a secondary concern. Yet the underlying assumption is that engineering and related technical/professional students and faculty, and their partners and employers from business, industry, government, and school systems, will contribute more significantly than anyone else to these and other high-tech advances. This assumption also pervades the critique of higher education for selling its soul to high-tech corporate America, as articulated by, among others, historian David F. Noble.[41] The Commission's January 1996 charge to its members reminds them that Justin Morrill "intended our institutions to provide a practical education, not a classical ideal," and to embrace the "'industrial class'—ordinary laborers, farmhands, workers, and their children."[42]

What, however, constitutes a "practical education" in our contemporary student-as-consumer culture may pose further dilemmas for effecting the Commission's recommendations. It is one thing for engineering in its conventional forms to blend with agriculture, computer science, chemistry, biology, and medicine to form such new fields as space engineering, geochemistry, biotechnology, environmental engineering, and space engineering.[43] These obviously constitute "legitimate" subjects for the broadened notion of "scholarship," for "integrated learning," and for "engagement." But it is quite another thing for institutions that wish to distinguish themselves from vocational schools, community colleges, and the like to embrace less professional and more purely practical subjects.

Back in 1957, Eddy clarified the differences for his day:

> The Land-Grant Colleges have developed from institutions which were little more than trade schools. In this development, what was originally vocational education with emphasis on occupations has become professional education with the goal of broad training to fit a number of life careers. The colleges are not preparing plumbers and mechanics but engineers; not cooks and seamstresses but home economists; not so much practical farmers on the land as agricultural scientists. To do this, they have attempted to stress the fundamental disciplines above the practical techniques, the sustained pursuit of scholarship above the vocational art, and social consciousness above the narrow concern for employment and self-preservation. To them, social progress depends upon the highest degree of professional training.[44]

One wonders if such distinctions could be so easily maintained in our time, with those combined pressures from college students and top administrators and from business, industry, government, and school systems to be as "relevant" to the contemporary "real world" as any 1960s college left-wing protester could ever have dreamed.

The most the Commission says about "mission differentiation" is as follows:

> We must encourage state systems of public higher education to differentiate institutional missions in higher education so that resources are used in the most effective and efficient ways possible, from specific-job skills training to ongoing education not linked to special occupational skills.[45]

This does not strike me as sufficient.

In Maine, as in many other states, what currently wins legislative and public support in funding higher education is what creates jobs, especially high-paying jobs in technical fields. It is no accident that where all recent bond referendums on behalf of Maine's technical college system have passed overwhelmingly, none of the recent bond referendums on behalf of the University of Maine system have received such a large approval percentage, and some have outright failed. The technical colleges' slogan—"Putting Education to Work for Maine"—is hard to fault (or surpass). Not surprisingly, some of the courses given by the technical colleges are slowly but surely appearing in some of the University of Maine system offerings (and vice versa, to take advantage of the technical colleges' current popularity and their desire to broaden their own offerings).[46]

The Commission does acknowledge a significant, if barely outlined, role for the "arts and humanities" but rightly concedes that "the tension between liberal learning and technical education"[47] goes back to the land-grant movement's earliest days. Here, for once, the historical context is valid. The Commission contends that

...liberal learning helps prepare students to be better thinkers, better communicators, and more ethical and civic-minded members of the community.... Even if we believe the central purpose of the university is to assist individuals and communities to apply knowledge to the problems of their everyday personal and community lives, it follows that the arts and humanities form an essential core in every university. In all public institutions (and perhaps especially in land-grant institutions), with their special duty to serve society, it is as important to maintain the perspectives of the liberal arts as a basis for responsible action as it is to provide excellent technical education.[48]

Missing, however, from the Commission's materials is any more specific defense of the liberal arts. Any humanist who has ever defended the importance of history—or literature or philosophy or art, for example—alongside more (immediately) remunerative "technical education" courses would have welcomed much more of the above. For that matter, the Commission's related defense of "basic" research, or "knowledge for its own sake," is equally weak if well-intended.[49] Yet it is precisely knowledge for its own sake that finally distinguishes the genuine university from the technical or vocational school or the for-profit institution.

Nevertheless, the Commission's report *Returning to Our Roots: A Learning Society* should, if implemented, provide endless teaching, research, and employment opportunities for engineering students and faculty well into the twenty-first century. The report advocates using information technologies, "including new interactive, multi-media technologies" for enriching—and customizing—learning. Instruction can now be tailored to "societal, organizational, and individual needs." No less important, however, is the prompt creation of the "high-technology infrastructure that is required for access to information, for cutting-edge research collaboration, for better direct service to all of our constituents, and for teaching our students to evaluate, use, and shape information with the tools of technology." Indeed, building that infrastructure is deemed as crucial to the nation's future as "the construction of our railroads in the 19th century or our national highway system in the 20th century."[50]

Still, one misses in the Commission materials any reference to the historic association of land-grant institutions with expanding engineering education in the post–Civil War era, when so few other engineering schools existed. Likewise, one misses any reference to the historic association of land-grant institutions with the professionalization of engineering and the transition from so-called "shop culture" to "school culture." And, finally, one misses any reference to the historic association of land-grant institutions with engineering as embodying national goals and with engineering students and faculty as national leaders. Times have certainly changed, but the absence once again of greater historical context is disappointing.

Not only are engineering and engineering education not substantially addressed, but engineering is implicitly—and wrongly—conceived as applied science rather than as a distinct form of knowledge in its own right. For example, the Commission's January 1996 charge to its members notes Morrill's determination "to extend the riches of science for the benefit of all."[51] This may have been common parlance in the mid-nineteenth century, yet MIT—with its literal use of "technology"—opened its doors in 1865, and, like other nineteenth-century "institutes of technology," never conceived itself as merely applying scientific discoveries. Once more, when the Commission invokes the historical record, it is, at most, skimming the surface.[52]

Yet engineering is, in less explicit forms, crucial to the Commission's agenda, and in more than purely technical fashion. For the Commission really envisions not just a wholesale transformation of conventional campuses and extension services, as already outlined, but also a de facto social engineering project that would constitute "Returning to Our Roots" on a grand scale. Land-grant and state colleges and universities "must again become the transformational institutions they were intended to be."[53] The project is the application of high-tech advances to: a) democratize access to higher education; b) promote the ever greater "diversity" of students, faculty, and staff; and c) provide "access to success," or "access to the full promise of American life."[54] The traditional linkage of the land-grant movement with expanding democracy—as epitomized in such works as Ross's *Democracy's College*, Eddy's *Colleges for Our Land and Time*, Nevins's *The State Universities and Democracy* (1962), and Joseph Edmond's *The Magnificent Charter* (1978)—is repeated in the Commission's reports and documents. Moreover, the sense of the inevitability of greater democratization in higher education that pervades all four of those books is akin to the sense of inevitability of greater technological transformation benefiting all that pervades the Commission. Nevins, in fact, describes the growing faith in inevitable technological progress in the mid-nineteenth century as spilling over into that sense of inevitability of greater democratization in higher education. "Anything seemed possible to the nineteenth-century civilization that was conquering one land after another by the industrial revolution and a new social enlightenment."[55] But the Commission overlooks the possible negative consequences for democracy and diversity of the very high-tech advances that lessen the appeal of the traditional classroom and campus experience and so lessen the old-fashioned notions of greater democracy and diversity stemming from the intellectual and social interactions of students, faculty, staff, and community members in "real-time" settings.

Furthermore, there is a considerable difference between what the Commission characterizes as Morrill's desire to prepare graduates for "the 'profession of life' itself" and the contemporary American Dream that the Commission describes as "the opportunity represented by modern technology and the development of a 'wired nation' practically overnight."[56] The latter does not appear to embrace non-entrepreneurial activities on the part of NASULGC institutions and their prospective partners as much as in earlier times. True, Ross observes that on the early land-grant campuses dissenters from the belief in "business enterprise" and in other material aspects of the "gilded age"[57] were met by strong opposition. But the belief in education itself as the solution to society's ills was supreme, whereas, by contrast, it is increasingly difficult today to distinguish placating "business enterprise" from reforming higher education. The most explicit such statement comes from John Byrne, executive director of the Kellogg Commission: "What we are talking about is individual access to success *through* higher education—not simply access *to* higher education—[but] access to the full promise of American life."[58] The ultimate measure of "success" here would appear to be quantifiable: material, financial, occupational, and so forth. The "social consciousness" that Eddy describes as characterizing much of the land-grant research efforts in the first century or so—"efforts which have, as their ultimate purpose, the good of humanity"—is certainly not stressed in the Commission's reports and documents. Nor is the "training for democratic citizenship"[59] that, according to Eddy, also characterized land-grant institutions in earlier days. Eddy likely romanticizes the past, but not, I suspect, enough to erase these differences in vision between the past and the present.

Indeed, if, according to the Commission, the NASULGC institutions can provide such unprecedented access to higher education and in turn to the American Dream, plus such unprecedented diversity, the nation will be on the road to "something quite new": "A Learning Society." This is the finale of the social engineering project but is not defined by anything beyond the same points made in other Commission materials and, for that matter, in many other forums outside of the Commission. For example, a "Learning Society" "stimulates the creation of new knowledge through research and other means of discovery and uses that knowledge for the benefit of society." Not quite an original notion. For the "Returning to Our Roots" historical context, we get only the following:

> The new role before us is no less important than the one we assumed under the original Morrill Act of 1862, or the Hatch Act of 1887, or the 1914 Smith-Lever Act. The first called on us to democratize higher education. The second asked us to focus research on technological advances in agriculture and industry. And the third made public service and engagement guiding principles in our work.[60]

Surely earlier versions of a "Learning Society" characterized much of American society from colonial times on, but especially beginning with the early-nineteenth-century mechanics institutes, lyceums, public schools, and public colleges and universities; plus the post–Civil War short courses for part-time students in early land-grant institutions themselves and, of course, those institutions' own farms and shops for teaching and research. If "lifelong learning" in a literal sense may be a newer phenomenon, versions of even it began long before the 1990s, as the Commission more or less concedes.[61]

In any case, in light of the Commission's commitment to the two-sided "engagement" of NASULGC institutions with their external partners, this version of social engineering would not remotely resemble traditional top-down directives to those outside of the academy. Quite the opposite: "Our institutions must become *genuine learning communities*" in which no one, least of all faculty, appear as anything but de facto peers, as fellow learners. Such "learning communities," not surprisingly, "should be *student centered*, committed to excellence in teaching and to meeting the legitimate needs of learners, wherever they are, whatever they need, whenever they need it."[62] To be sure, top-down directives from largely non-teaching administrators will, as with other Commission recommendations, ultimately impose a sense of equality on the part of everyone else should any dissent arise from faculty or others. As the Commission puts it:

> . . . our institutions must act in concert if they are to succeed. We very much doubt that the land-grant movement could have transformed American higher education in the 19th century if our institutions had been created piecemeal. Nor could our 20th-century success in broadening access and building research capacity have been developed one-institution-at-a-time. The true power of the land-grant movement manifests itself when we join hands and move forward together.[63]

Finally, the Kellogg Commission, like nearly everyone else today, pronounces the emergence of a high-tech-based global economy that must be recognized, respected, and obeyed. (The one exception to this is revealing: criticism of faculty for excessive loyalty toward "national and international societies and associations of their disciplinary peers" and insufficient loyalty toward "their immediate colleagues, students, and institutions."[64]) Since the "Returning to Our Roots" theme does not, as historical analysis, extend beyond the surface of post–Civil War American history, any notion of earlier global economies having been in existence for thousands of years is naturally absent. Yet the current global economy bears on the future of the land-grant ethos in more ways than economics and technology alone. For if, as we are repeatedly told, the global economy is a homogenizing force among nations—for better or for worse—it is probably no accident that public higher education is treated as a homogeneous entity throughout the Kellogg Commission's reports and documents. The quotation above about the need to "act in concert" exemplifies this outlook. True, any such ad hoc body as the Commission must make generalizations in order to drive home its fundamental message. But one still misses any sense in its materials of the uniqueness of each of the 203 institutions under the NASULGC umbrella, thanks to varied community, state, and regional cultures, economies, politics, traditions, and values. Once more, the technological determinism that pervades the Commission's analyses is decisive: there is no recognition (whatsoever) of the extraordinary degree to which culture, broadly defined, shapes technology at least as much as technology shapes culture. Indeed, the official land-grants are lumped with other public colleges and universities and are collectively treated as one.

No one can deny that many of the 203 NASULGC institutions are more similar than different, but one misses any sense of the differences between a small land-grant like my institution, the University of Maine, and a large one like Iowa State. To be sure, in reviewing the latest strategic plans of those two schools, one finds endless similarities not just in goals but also in rhetoric—not least, the familiar phrases of the Kellogg Commission itself.[65] But the University of Maine and Iowa State are hardly alike, while the seventeen historically black member institutions of NASULGC and the twenty-nine Native American tribal colleges have very different agendas from both Maine and Iowa State. In fact, the tribal colleges resemble the original land-grants in their focus on vocational training, especially for jobs on often impoverished reservations.[66] Paradoxically, a commission that celebrates "diversity" when it comes to race, gender, religion, and class apparently has no room for appreciating diversity here. However romanticized the histories of the land-grant institutions provided by Ross, Eddy, Nevins, and Edmond, all four paid attention to differences as well as similarities among the original land-grant institutions. Consequently, "Returning to Our Roots" ought also to mean becoming more acquainted with the rich and complex histories of land-grant and other public universities.

The 1997 inaugural address of President Hoff of the University of Maine, much of it implicitly derived from Kellogg Commission discussions, analyses, and rhetoric, is nevertheless clearer than any official Commission document about reengineering the land-grant university.

> As we envision how to position the University of Maine for the coming years, we need to look no further than the Land Grant roots and tradition of the university. By being true to our roots, we can reorient the university to make it the university of the future. How can this apparent paradox be true? Very simple: access and engagement—the very things that made us what we are—are the keys to a vital future. Our challenge is to redefine access and engagement in ways that address the current and future needs of Maine.[67]

As Eddy concluded in 1957 about the land-grant colleges and universities, "Americans thus have reflected their values in the development of these educational instruments."[68] The same point, of course, applies to the Kellogg Commission. The Commission reflects the pervasive contemporary faith both in high tech as the alleged shaping force of American life—not least, public higher education—and in high-tech advances as virtual panaceas to the various educational and social problems it addresses. The dilemmas created by that almost blind faith—above all, the future of traditional campuses and of traditional academic instruction—are minimized by the Commission, just as they are in the marketplace, as with the dangers to children and other innocents in Internet chat rooms, for example, or the (in)appropriate uses of cell phones in driving and in natural settings. Dissenters from that faith risk being dismissed as twenty-first-century Luddites.

Invoking its often simplified version of the educational past in order to justify the present and especially the future, the Commission offers a vision that at once is politically appealing to the general public, to government at all levels, to K–12 education systems, and to corporate America, and politically correct to others who also demand reform. "Returning to Our Roots" and "Renewing the Covenant" are attractive phrases, as are "Learning, Discovery, and Engagement," "Learning Society," and "Lifelong Learning." Yet those at the top of the higher educational establishment show no evidence of wishing to reduce an elitist status that is nevertheless at odds with the ethos of democratization they would impose everywhere else. Their social engineering project, as outlined above, may thus fall victim to the criticisms of other social engineering projects in higher education—for instance, the affirmative action and diversity efforts of the 1980s and 1990s in student, faculty, and staff recruitment and retention—as being intended as much to assuage the guilt of liberal educational elitists as to bring about genuine change.

Ironically, though, the more serious challenge to the moral authority of the Kellogg Commission comes from contemporary technology itself: the sense of "empowerment" felt by ever more technically proficient citizens—thanks to the very high-tech advances hailed by the Commission—and the consequent need to rely ever less on technical experts. The vaunted ability of computers, e-mail, web sites, cell phones, fax machines, and so forth to allow individuals not merely physical and geographical but also psychological and ethical distance from otherwise prevailing moral—and even legal—authority undermines the Commission's social engineering project. For that project presumes an ability to forge a nationwide consensus based on active participation in educational enterprises, be it on campus or off. Newly empowered students, for example—who are encouraged to take courses and earn degrees at their utter convenience, and often alone—are hardly akin to earlier generations of students in pioneering land-grant institutions, who saw themselves as reformers expanding and democratizing American higher education by their sheer visibility (rather than, as increasingly today, by their invisibility behind computer screens and the like). It is no accident that "Total Quality Management," the epitome of individual empowerment as applied to organizational structures and operations, has by now found a home at American colleges and universities decades after becoming part of offices, factories, and other workplace locales. "Strategic planning," "active listening," "team development," "conflict resolution," "customer service," and other ingredients of TQM may have already functioned successfully at some levels of college and university administration, but TQM's supreme faith in individuals as "experts on their own work" may prove more difficult.[69]

More than capitulation to the marketplace in the name of profits and efficiency, or capitulation to postmodern modes of thought that question the meaning and value of truth and objectivity,[70] high-tech advances, along with the extraordinary sense of individual empowerment that they both encourage and reflect, threaten the Kellogg Commission's ambitious enterprise and in turn the future of the traditional land-grant university. Historians of technology who routinely emphasize the unexpected consequences of technological advances, and so their often profoundly mixed blessings, would not be surprised if the Commission's enterprise did not succeed for this reason. But they might be amused at the supreme irony of it all.

Notes

1. See Nathan M. Pusey, *The Age of the Scholar: Observations on Education in a Troubled Decade* (Cambridge, MA: Harvard University Press, 1963).

2. "Announcement of Kellogg Commission: New Commission to Bring Reform to State and Land-Grant Universities Funded by Kellogg Commission," Press Release, January 30, 1996, 1.

3. Ibid., 3.

4. E. Gordon Gee, quoted in ibid., 2.

5. Kellogg Commission, "Taking Charge of Change," brochure, June 1996, cover letter.

6. On this proposed Act, see note 56 below.

7. See, for example, Paul W. Gates, "The Morrill Act and Early Agricultural Science," *Michigan History* 46 (December 1962), 289–302, esp. 300–302, regarding legislative and other conflicts over what became the Morrill Act and the further conflicts over where to locate the new land-grant institutions and over the demands of those who didn't get them to use land grant-funds for non-land grant colleges, including religious ones. See also Scott Key, "Economics or Education: The Establishment of American Land-Grant Universities," *Journal of Higher Education* 67 (March–April 1996), 196–220, which argues that the Morrill Act was more concerned with economics than with education.

8. Allan Nevins, *The Origins of the Land-Grant Colleges and States Universities: A Brief Account of the Morrill Act of 1862 and Its Results* (Washington, DC: Civil War Centennial Commission, 1962), 26–27.

9. "Taking Charge of Change," cover letter.

10. Ibid., 3.

11. Edward D. Eddy, Jr., *Colleges for Our Land and Time: The Land-Grant Idea in American Education* (New York: Harper, 1957), 271.

12. James J. Stukel, "The Future of Land Grant Universities," speech delivered in Corpus Christi, Texas, February 19, 1998, 2, 9, 4, 6. The speech can be obtained through NASULGC's website, which is how I came upon it.

13. Ibid., 9.

14. Eddy, 285.

15. Ibid., 10.

16. See, for example, the proposed online Portland (Maine) College as outlined by Gordon Bonin, "Web College Envisioned for Maine: Degrees to Be Earned Online," *Bangor Daily News*, Saturday/Sunday, November 27–28, 1999, A1, A12. Ironically, because of questions raised by some Maine legislators over the funding of this enterprise, its mailing address—if not its offices, such as they are—has since been moved to rural Montana.

17. As quoted, for example, in the Kellogg Commission report *Renewing the Covenant: Learning, Discovery, and Engagement in a New Age and Different World*, March 2000, 1.

18. Ibid., 4.

19. Paula Evans Petrik, then a professor of history at the University of Maine, quoted in Scott Jaschik, "Historians Differ on Impact of Distance Education in Their Discipline," *The Chronicle of Higher Education* 46 (January 21, 2000), A43.

20. Laurence R. Veysey, *The Emergence of the American University* (Chicago: University of Chicago Press, 1965), 347. The rest of the paragraph on 347 is a good summary of Veblen's critique.

21. See Eddy, 283.

22. Big Bill Haywood, quoted in David Montgomery, *Workers' Control in America: Studies in the History of Work, Technology, and Labor Struggles* (New York: Cambridge University Press, 1979), 9.

23. See *Productivity: Productivity for the Academic World* (Portland, OR: Productivity Press, Summer 1998), 2 and cover. As the accompanying May 28, 1998, "Dear Educator" letter puts it, "Regardless of your academic specialty, the keyword in delivering an education that will serve us in the modern world is change."

24. Kellogg Commission, *Renewing the Covenant*, 7.

25. Kellogg Commission report, *Returning to Our Roots: Toward A Coherent Campus Culture*, January 2000, 8.

26. "Opinion: Do Land-Grant Schools Do Too Much Research?" *Feedstuffs* 64 (October 26, 1992), 8.

27. Earle D. Ross, *Democracy's College: The Land-Grant Movement in the Formative Stage* (Ames: Iowa State College Press, 1942), 137, 139.

28. Peter S. Hoff, "Back to the Future: The University of Maine and Its Land Grant/Sea Grant Tradition," inaugural address, November 21, 1997, 8, 7.

29. John V. Byrne, "The Public University of the Future," Kellogg Commission unpublished paper, January 25, 2000, 1. Byrne served as executive director of the Commission and was president of Oregon State University at the time of his appointment. His paper summarizes the Commission's conclusions and recommendations. Before the Commission ever met, the University of Maine's motto was "the state's center of learning, discovery, and service to the public." It had been devised in 1992 in order to communicate the University's contemporary mission to the general public. Ernest Boyer's work, cited below, had no bearing on this reformula-

tion. My thanks to John Diamond, then director of public affairs, for this information. I do not know how representative that motto is among public universities.

30. Eddy, 269, 271.

31. Kellogg Commission report, *Returning to Our Roots: The Student Experience*, April 1997, vi.

32. Ernest L. Boyer, *Scholarship Reconsidered: Priorities of the Professoriate* (Princeton, NJ: The Carnegie Foundation for the Advancement of Teaching, 1990), 75.

33. Kellogg Commission report, *Returning to Our Roots: The Engaged Institution*, October 1998, 17.

34. Boyer, 78–79.

35. Kellogg Commission, *Returning to Our Roots: The Engaged Institution*, 29.

36. Ibid., 30.

37. Kellogg Commission, *Returning to Our Roots: Toward A Coherent Campus Culture*, viii–ix.

38. Ibid., 15. On the "reform agenda of the 1990s," see ibid., 8.

39. Ibid., x.

40. See Robert W. Harrill, "Evolving Curricula in the New Century: Putting Universities Back in Touch—A Prototype Program that Links the Community and Institution," *Journal of College Science Teaching* 29 (May 2000), 401–407. Harrill's article is in part an endorsement and in part a critique of the Commission's recommendation for more interdisciplinary efforts, as outlined in *Returning to Our Roots: The Engaged Institution*, vii, 32.

41. See, for example, David F. Noble, "Selling Academe to the Technology Industry," *Thought and Action: The NEA Higher Education Journal* 14 (Spring 1998), 29–40. See also David L. Kirp, "The New U," *The Nation* 270 (April 17, 2000), 25–29, a review of two contrasting books on contemporary higher education, and Arthur Levine, "The Soul of a New University," Op Ed, *New York Times*, March 13, 2000, A25.

42. "Kellogg Presidents' Commission on the 21st Century State and Land-Grant University," Letter to Commission members, January 1996, 1.

43. I take these examples from Lawrence P. Grayson, *The Making of An Engineer: An Illustrated History of Engineering Education in the United States and Canada* (New York: Wiley, 1993), 262, still a most useful and comprehensive work. As Grayson noted, "These new fields, along with advances in materials, instrumentation, conceptual and analytical techniques, and methods of processing and displaying information and data, are forcing change in the content and scope of engineering education" (262). Interestingly, where agriculture at land-grant institutions has been the subject of somewhat defensive discussions in recent years, engineering at land-grants has not, to my knowledge, been subjected to such critiques. It is widely assumed to be both relevant and accessible. See, for example, M. L. Westendorf, R. G. Zimbelman, and C. E. Pray, "Science and Agriculture Policy at Land-Grant Institutions," *Journal of Animal Science* 73 (June 1995), 1628–38, and "Colleges of Agriculture at the Land-Grant Universities: Public Service and Public Policy," *Proceedings*

of the National Academy of Sciences of the United States 94 (March 4, 1997), 1610–11.

44. Eddy, 280.

45. Kellogg Commission report, *Returning to Our Roots: A Learning Society*, September 1999, 29.

46. In Maine, unlike most other states, the legislature conveniently passes on decisions on large-scale funding to the public, which therefore routinely votes on referenda on the funding, through bonds, of various projects, not just those in education.

47. Kellogg Commission, *Renewing the Covenant*, 9. Eddy, 283, makes a similar point.

48. Ibid.

49. See ibid., 10.

50. Kellogg Commission, *Returning to Our Roots: A Learning Society*, vii, 31.

51. "Kellogg Presidents' Commission on the 21st Century State and Land-Grant University," letter to Commission members, January 1996, 1. See also, for instance, Kellogg Commission, *Returning to Our Roots: A Learning Society*, 28. Admittedly, Eddy, for example, is no clearer; see 274 and 281. But Eddy, 282, does recognize the need to support both "pure" and "applied" research.

52. To be sure, the Kellogg Commission is hardly alone in our day in continuing to equate technology with applied science. See Howard P. Segal, "The Third Culture: C.P. Snow Revisited," *IEEE Technology and Society Magazine* 15 (Summer 1996), 29–32.

53. Kellogg Commission, *Returning to Our Roots: The Student Experience*, 1.

54. Kellogg Commission report, *Returning to Our Roots: Student Access*, May 1998, 2. Even this, however, does not satisfy Mary Burgan, general secretary of the American Association of University Professors—an organization not exactly known for its militant stands toward anything—because it lacks "the missionary zeal that affirmative [action] programs once embodied" (Burgan, "Access: A Matter of Justice," *Academe* 98 [July–August 1998], 72).

55. Nevins, 8.

56. "Kellogg Presidents' Commission on the 21st Century State and Land-Grant University," letter to Commission members, January 1996, 1; and Kellogg Commission, *Returning to Our Roots: Student Access*, 2.

57. Ross, 150.

58. Byrne, 3.

59. Eddy, 272, 281. Eddy also makes some suggestive comments about (1) the alleged prevailing sentiment in the first century of the land-grant movement that the "basic unit of the social idea is the individual person" (270); and (2) the alleged belief in those years that "the individual family farm has been the bulwark against the movement toward the large corporate agricultural enterprise" (283).

60. Kellogg Commission, *Returning to Our Roots: A Learning Society*, vii, 9. There *is* a modest elaboration on this, and a listing of more federal legislation crucial to the evolution of public higher education in *Renewing the Covenant*, 12–13, along with an outline of the proposed Higher Education Millennial Partnership Act, 12–15.

61. See ibid., 33.

62. Kellogg Commission, *Returning to Our Roots: The Student Experience*, v–vi (emphasis original).

63. Ibid., 2. On the Commission's attempt to avoid the "top-down corporate model" of academic leadership, see Kellogg Commission, *Returning to Our Roots: Toward A Coherent Campus Culture*, 9.

64. Kellogg Commission, *Returning to Our Roots: Toward A Coherent Campus Culture*, viii.

65. See the University of Maine's *BearWorks 2.0: An Action Plan: A List of Priorities, Goals, and Objectives Designed to Enable the University of Maine to Fulfill Its Mission and Potential as the Flagship Campus of the State University System* (1998), which is currently being expanded and refined; and *Becoming the Best Land-Grant University: Strategic Plan for 2000–2005: Pursuing Excellence as Iowa's Engaged University*, Iowa State University of Science and Technology (1999).

66. See Alvin L. Young, "Tribal Colleges: New Land-Grant Institutions Grow and Thrive in America," *Resource: Engineering and Technology for a Sustainable World* 5 (May 1998), 6–7; and, for a rather different assessment, Beth Daley, "On Reservations: A Failing Mission: Decades-Long Effort Has Been Unable to Help Tribal Colleges Thrive," *Boston Globe*, March 11, 2000, A1, A11. The latter is an excellent example of how an existing culture can promote or undermine technological advance, contrary to the Commission's technological determinism.

67. Hoff, "Back to the Future," 4. See also ibid., 6.

68. Eddy, 286.

69. For elaboration on this paragraph, see Alan I Marcus and Segal, *Technology in America: A Brief History*, 2nd ed. (Fort Worth, TX: Harcourt Brace, 1999), 334–36, 340–42. See also John Holusha's obituary of W. Edwards Deming, the foremost advocate of TQM, in the *New York Times*, December 21, 1993, B7.

70. See David D. Cooper, "Academic Professionalism and the Betrayal of the Land-Grant Tradition," *American Behavioral Scientist* 42 (February 1999), 776–85.

■   Reinventing the Wheel

The Continuous Development of Engineering Education in the Twentieth Century

Bruce Seely

> "This school is designed to make good practical engineers."
> —*Engineering Education in a Land-Grant Context*

For most of the nineteenth century, a young man seeking to become an engineer (for American engineering was almost exclusively a male profession) served an apprenticeship, often related to railroad construction or in a machine shop. Only in the last third of the century was college generally accepted as the best preparation for an engineering career, and the nature and content of formal education varied widely from school to school. Indeed, American engineering schools have been characterized more by diversity than uniformity. Even before the Civil War, one might choose among the science schools at prestigious institutions (Harvard, Yale, Dartmouth), formal engineering programs at a number of universities (Georgia, Michigan, Rochester), and the first polytechnics (RPI, Brooklyn). After 1865, technical institutes (Worcester Polytechnic Institute, Stevens Institute) that combined shop apprenticeships and book learning were established.[1]

One of the most important developments in this history of the training of American engineers was the Morrill Act in 1862, and the land-grant universities it helped create.[2] Land-grant colleges have been celebrated as democratic institutions focused upon serving society. One of those services has been to educate large numbers of engineers. Indeed, by 1900, the majority of American engineers graduated from land-grant colleges, as 3,398 of the 4,459 degrees in mechanical engineering awarded that year were at land-grant schools. Similarly, 1,964 of 3,140 civil engineering degrees, 1,617 of 2,555 degrees in electrical engineering, and 822 of 1,261 graduates in mining earned degrees from land-grant institutions; the ten largest engineering colleges in the country were land-grant colleges.[3]

Apart from the numbers of graduates, land-grant colleges were important for helping develop two significant elements of American engineering education. First, they helped resolve the difficult debate concerning two competing aspects of engineering education: the role of theory (basic principles) and practice (hands-on experience). This pivotal issue aroused much passion among engineers after 1865.[4] Land-grant engineering schools were not alone in resolving this debate in favor of college training, but their solution "worked" for more than half a century. Second, land-grant schools provided the first homes for engineering research, but with an approach that differed substantially from the focus that characterized German-style research universities. Changing expectations and changing patrons eventually altered the land-grant approach to both education and research in the middle of the twentieth century. Yet since 1980, the educational style developed in land-grant colleges before 1900 has again seemed relevant to engineering educators.

Educational Patterns: 1870–1940

When the first land-grant colleges began opening their doors around 1870, no one would have confused them with Harvard. Historian Monte Calvert noted that until 1880, technical education at land-grant colleges was "hastily conceived and implemented. . . . The universities themselves were anxious to get land-grant money, but administrators frequently lacked faith in the value and future of technical education. As a result, the programs were given little attention."[5] The Texas Agricultural and Mechanical College in College Station, for example, admitted students with a ninth-grade preparation, and a number of programs focused on manual skill development in such areas as carpentry, cabinetmaking, and black-smithing.[6]

Yet it was not long before a land-grant style began to emerge, centered upon providing an education immediately useful to people and a nation living through industrial and agricultural revolutions. Thus machine shop practice was emphasized at most schools, while every engineering student learned to use a surveyor's transit. Most classes had little theoretical content, but perhaps for that reason engineering and technical courses quickly proved more popular than agriculture programs. As the University of Illinois summarized its educational goals in its 1872 engineering college catalog, "This school is designed to make good practical engineers. . . ."[7]

It fell to faculty members like Stillman W. Robinson, the first and for a time only engineering professor at Illinois, to meet that goal. Robinson grew up on a Vermont farm and apprenticed in a machine shop before earning a civil engineering degree at the University of Michigan in 1863. He worked for five years as an assistant on the U.S. Lake Survey, and also taught mining engineering and geodesy at Michigan from 1866 to 1869. In January 1870, he arrived in Urbana at the Illinois Industrial University. The courses offered that year were Mechanic Science and Art, Civil Engineering, Mining Engineering and Metallurgy, and Architecture and Fine Arts.[89]

Robinson typified early engineering instructors. Like most, he taught everything from physics and geodesy to mechanical and mining engineering. Equally typical was his effort to combine practice with systematic classroom instruction. His career reflected this approach. Robinson the college teacher published papers on structural mechanics and magnetic circuit breakers; he inaugurated laboratory work in physics in 1875 and the first courses in materials and hydraulics. Robinson the practical engineer devised and his students built apparatuses for classroom demonstrations in materials courses. He also helped build a steam engine for the campus machine shop and the tower clock for the union. He patented a rock drill for the Hoosac Tunnel and devised a lawn mower and sewing machine. In 1878, he became dean of engineering, but moved to Ohio State nine months later for $450 more a year. He remained in Columbus until 1895, when he decided to pursue interests in shoemaking machinery and amassed a fortune.

Robinson and faculty like him were above all teachers who prepared students for real-world engineering. They found educational opportunities quite literally in building the campuses on which they worked. At Iowa State, for example, Anson Marston designed the water tower that still dominates the campus, and he, too, constructed lab equipment.[10] The laboratories that used that equipment were an important element in the success of the land-grant engineering schools, for the hands-on laboratory work was a primary response to the belief of traditionalist engineers that only exposure to shop work (in mechanical engineering) and field training (in civil engineering) adequately prepared young engineers. Like almost all engineering colleges during the last third of the century, land-grant schools adopted shop elements in their curricula, borrowing the idea from the technical institutes, which combined a version of the traditional machine shop apprenticeship with formal courses. But unlike the institutes, land-grant schools subordinated shop work to classroom exercises.[11]

Still, there was never a doubt that engineers were expected to operate in the real world, to solve real problems, and to know the tools of their trade. As the Department of Mechanical Engineering at Illinois noted in its statement of purpose in 1873, "The instruction, while severely scientific, is thoroughly practical, aiming at a clear understanding and mastery of all mechanical principles and devices."[12] Not surprisingly, the faculty almost universally had worked as engineers before returning to teach. Many moved back and forth between business and colleges during their careers. Few had doctorates, for a bachelor's degree, tempered with real experience, was the credential that defined this generation of engineering faculty. In the end, men like Stillman made a home for engineering in colleges by developing a balanced education that included both hands-on experience and attention to basic principles.

Over the last decades of the nineteenth century, land-grant engineering schools grew enormously in popularity. While Robinson had started alone, by 1885, Illinois had four professors, two assistants, and two instructors. A decade later, the College of Engineering at Illinois employed twenty-five teachers, and by 1905, there were forty-three. These years also saw substantial changes in the structure of the schools, as departments took shape around the traditional fields of civil, mechanical, and mining engineering, as well as new disciplines, such as metallurgy, electrical, and chemical engineering. But the basic style of education did not fundamentally change until the beginning of World War II. There were, to be sure, adjustments to curricula to take advantage of new knowledge and to incorporate new subfields of engineering. And in certain fields, especially the science-based engineering domains of electricity and chemicals, hands-on experience and rules of thumb increasingly failed to provide adequate guides for innovation or management of complicated technical systems. Even so, the tendency within the land-grant schools—and most other American engineering colleges as well—was to maintain the traditional orientation toward practice. Harris Ryan, a leading electrical engineer with a distinguished career at Cornell and Stanford, succinctly summarized the basic assumption: "The spirit of engineering can not be acquired through academic life."[13]

A number of reformers, however, complained about the poor preparation of American engineering students entering the newer "high-tech" fields. Preeminent among them was mechanical engineer Robert H. Thurston. A graduate of Brown University in philosophy and engineering, Thurston taught at the Naval Academy in Annapolis in the 1860s before moving to Stevens Institute in 1871. Then he became dean of Cornell University's Sibley College of Mechanical Engineering in 1885 and set out to introduce a more scientific engineering curriculum at Cornell.[14] According to Monte Calvert, Thurston had concluded that "engineering education should be primarily a blend of French and German ideas. From France came an emphasis upon math and science and the concept of the high-level professional school; from Germany came the practice of setting up schools to train technical personnel at all levels, with the research institution at the top." Cornell's high entrance standards gave Thurston students prepared for serious study of the mathematical and scientific principles, and allowed him to replace some hours in machine shops with time for "calculations" and basic science courses.[15]

Yet almost no other American engineering school, land-grant or otherwise, followed Cornell's lead. By 1919, Maurice Caullery, a French observer who toured American universities, reported, "There is nothing in the United States comparable to the preparation in our courses of the *École Polytechnique* or the *École Centrale*. The first-year students, the freshmen, of the engineering schools, are very weak. It is none the less true that the American engineer gives abundant proof of all the qualities which are expected of him. . . . He is first of all a man of action."[16] Complaints continued to be heard, including in a mid-1920s survey of engineers.

"The American is lacking in a good solid grounding in the elements of engineering mechanics, physics, chemistry, and the natural laws by which the world goes around," wrote one engineer. "It is primarily the duty of the faculties of our American universities to eliminate as far as possible the shop courses, the so-called research courses, and in fact, all so-called practical work and concentrate all effort in preparing the foundation to better advantage."[17]

After 1920, European-born engineers such as Theodore von Kármán and Stephen Timoshenko began introducing approaches to engineering problem solving that rested upon rigorous mathematical analysis. But the land-grant schools generally retained a strong emphasis upon balanced courses and practical laboratory courses, with a bias toward hands-on experience. Thus Embury A. Hitchcock, a professor of experimental engineering at the Ohio State University, introduced his mechanical engineering students to machinery, design, and practical problem-solving by assigning senior projects such as calculating heat balances for moving locomotives on the Hocking Valley Railroad or for the Columbus Water Works steam pumping engines. Students at Cornell tested street railway motors and generators in Buffalo, while textile engineering students at Georgia Tech ran a factory.[18] In short, the land-grant view of a balanced engineering education lasted well into the middle of the twentieth century.

Land-Grant Research and Engineering Experiment Stations, 1900–1940

If there was little change in the educational approach of the land-grant schools through the early twentieth century, that did not mean there were no changes in land-grant engineering programs. The most important adjustment was the way that land-grant engineering colleges introduced engineering research into an academic setting. This development was directly connected to the land-grant philosophy of service that had taken root on the agricultural side of many schools. The basic conception was to carry education beyond the classroom, in the form of short courses of immediate practical utility and extension programs. Both became central activities at land-grant colleges, and the problem-solving ethos undergirding them also grew to encompass systematic and eventually scientific investigations of crucial problems facing local farmers. The establishment of agricultural experiment stations in California (1873), Connecticut (1875), North Carolina (1877), and New York (1878) marked this development. By 1886, nineteen states had formed agricultural research stations, and a year later the federal government began providing $15,000 annually to every state to support them. The stations studied everything from animal and plant diseases to new varieties and breeds and improved production techniques and mechanization. In the process, they demonstrated that science and research could be harnessed for utilitarian ends, even if some problems required years of study into fundamental scientific questions.[19]

Above all, the stations established practically oriented research as a central activity in the land-grant schools.

This orientation of service to society came to be known as the Wisconsin Idea, after that premier land-grant institution formally addressed these activities as part of its mission, arguing that "the acquisition of knowledge carries with it the responsibility to apply that knowledge to benefit society."[20] This logic also came to guide engineering activities, as the style of teaching showed. But engineering faculty could emulate their agricultural colleagues only slowly in terms of research. Cornell's Robert Thurston was almost alone in pressing for engineering faculty to routinely engage in serious research, although a number of his students made research part of their professional activities. Anson Marston, who became dean of engineering at Iowa State, was perhaps the most successful in this regard.[21]

After 1900, however, research began to become more evident in land-grant engineering colleges. Some faculty began while consulting for industry. Electrical engineer Dugald Jackson, of Wisconsin and later MIT, believed consulting contacts would attract industrial funds to the colleges. One educational benefit was the way that faculty consultants were exposed to the most current problems in industry, enriching their teaching.[22] Other faculty simply made time for research. Arthur Newell Talbot, for example, taught civil engineering at Illinois after 1885, but also conducted studies on several topics in municipal engineering and materials behavior. He published a formula for the flow of water in culverts in 1888, and another on rainfall and runoff. He developed a method for laying out railway curves in 1891, and in the 1890s studied septic tanks, sewage disposal techniques, culvert design, and highway pavements and materials, which led him to study the strength of brick and concrete.[23] At Purdue, mechanical engineer W. F. M. Goss purchased a steam locomotive in 1891 and "designed and installed the locomotive testing plant at Purdue, the first of its kind in the world; and with this equipment he conducted numerous tests which materially added to the knowledge of locomotive performance."[24] At Ohio State, Edward Orton, Jr., launched a program in ceramic engineering education in the 1890s that rested on close links to the state's ceramic industry.[25] Several civil engineers, including Marston at Iowa State, studied brick, cement, pipe, road-building materials, and sewage treatment in the 1890s. But these were exceptional individuals, all of them pioneers in their fields.

In 1903, research efforts received a huge boost with the formation of the first engineering experiment station at Illinois; a second was set up at Iowa State in early 1904. The stations institutionalized research efforts, paralleling the agricultural stations even though federal funding for the engineering stations never appeared, despite high hopes.[26] The most important common pattern between the stations was their service missions, for the engineering stations were to serve urban and industrial society in the same way that the agricultural stations helped farmers. This logic was a perfect reflection of the Progressive mentality and its fascination with efficiency, trusted experts, and action based on information.

Anson Marston at Iowa State College was the leading advocate of this pub-lic-service approach to research. He believed engineering schools should help solve the problems facing an urban, industrial American, both through trained graduates and through research. An archetypal Progressive-era figure, Marston told the Society for the Promotion of Engineering Education in 1900 that "a tech-nical school would be one of the most potent agencies imaginable for the better-ment of the welfare of the people, and for the progress of modern civilization."[27]

This vision was shared by the University of Illinois, and the two schools gave their stations almost identical charters. Thus the purpose of the Illinois station was

> ... to carry out investigations along various lines of engineering and to study problems of importance to professional engineers and to the manufacturing, railway, constructional, and industrial interests of the state. . . . It is believed that this experimental work will result in contributions of value to engineering sci-ence and that the presence of such investigations will give inspiration to stu-dents and add efficiency to the College of Engineering.[28]

Or as the Iowa Station's charter put it, the station should "assist the urban popula-tion in solving the technical problems of urban life [and] the engineering prob-lems of the agricultural population."[29]

Practical investigations dominated the agendas of both stations until 1940. At Iowa, researchers improved sewage treatment, land drainage, and highway con-struction. Engineers at Illinois also tackled these problems and examined materials used by local governments, with an eye to setting standards. Talbot was the most in-volved faculty researcher, as he produced thirteen bulletins and co-authored nine more on reinforced concrete and optimal mixes of water, sand, aggregate, and ce-ment; on the performance of beams, slabs, columns, and frames fabricated from this material; and on railway engineering questions related to stresses in railroad rails and the performance of the ballast, ties, rail and hardware system.

Other land-grant colleges soon followed the pioneers, with stations appear-ing at Kansas State and Penn State (1912), Wisconsin and Texas A&M (1914), Maine (1915), and Colorado State College and Purdue (1917). By the end of the 1920s, thirty-one stations were in place. Universally, they repeated the pattern of Iowa and Illinois. Most had meager resources, with the exception of the Illinois station, which enjoyed a regular appropriation from the legislature. Texas A&M, on the other hand, survived on $2,000 to $3,000 annually through the mid-1920s; the station at Pullman, Washington, received $5,000 annually from the university until 1925. Other stations—Idaho, Wyoming, and Colorado—had no separate funds. But those that received state funds were expected to solve the technical problems facing governments.[30]

The other element in the research agendas of the early stations was service to "the manufacturing, railway, constructional, and industrial interests of the state."

In 1903, remember, systematic investigations by scientists and engineers were only just beginning within the largest corporations, so station promoters were in advance of the views of many in industry about the value of research. Their intention was to serve small industries that could not afford their own laboratories. Doing so, station engineers argued, would promote economic growth in their states. Marston pushed this argument hard, leading Iowa State engineers to assist firms developing Iowa's mineral resources: clay, brick, coal, and shale products. Illinois researchers attacked problems limiting the use of Illinois coal and conducted performance tests on many materials. The stations soon were conducting efficiency studies of electric generating plants, while Illinois hired W. F. M. Goss to build a locomotive testing plant to benefit the state's manufacturers of railroad equipment.

Both schools, however, placed restrictions on contact with industry. As the director of the station at Illinois explained, "The Station is conducted as an institution of scientific research rather than as a commercial testing laboratory. . . . No researches are undertaken with the object of obtaining information of chief value to some individual or company."[31] Under the public service ideal, the schools did not think it appropriate to use public funds to benefit individual firms. Rather, the stations should serve society by widely disseminating impartial research results to the entire industry. In 1914, Talbot amplified this point, noting that college laboratories "enjoy the very full confidence of the public who feel that these laboratories are not connected with special interests and that research problems will be handled with judicial fairness and impartiality."[32] The stations were an ideal example of the Progressive faith in experts.

After 1920, however, the stations began to expand their work for industry, in light of the explosion of interest in industrial research, especially in larger corporations, which had established 1,625 laboratories by 1930, up from 300 at the start of the decade. The Illinois station signed its first cooperative research contract in 1916, with the Association of Manufacturers of Chilled Car Wheels. Other projects followed, including home furnace, boiler, and radiator testing for the National Warm Air and Ventilating Association; studies of fatigue of metals for the Engineering Foundation; numerous studies of porcelain, electric cables, refractories, and other subjects for the Utilities Research Committee of Chicago, the National Brick Manufacturers Association, the Illinois Master Plumbers' Association, the Clay Products Association, and other trade groups. By 1928, sponsored projects generated $100,000 for the Illinois station.[33]

Other stations also sought industrial contracts during the 1920s. Ohio State engineers reinforced connections to Ohio's ceramic industry, building a facility to test glazes and clays for the Ohio Ceramic Industries Association and other trade groups. Engineers at Kansas State University studied rural electrification, while research fellowships from corporations and trade associations supported similar projects at almost every station. Researchers studied culvert pipe at Iowa State for the American Concrete Institute, the American Railway Engineering Association, the American Concrete Pipe Association, and various professional engineering groups.[34] But Purdue's dean of engineering, A. A. Potter, built his school's experiment station into the country's largest. In 1922, cooperative projects for the American Railway Association and the Indiana Quarrymen's Association brought only $8,750 to Purdue, but such work increased after a 1926 conference on industrial research. Potter announced that he wanted Purdue "to become the Indiana Bureau of Standards and Indiana's Mellon Institute," as well as "the technical laboratory of our smaller industries." He argued, "Scientific research can make the state of Indiana the type of commonwealth we desire it to be. New discoveries will make new industries, and greater prosperity will help us to induce a greater number of engineering graduates to stay here in Indiana."[35] Sponsored projects jumped from $127,000 in 1926 to $275,000 in 1929—more than half of the outside research funds at all stations combined. Researchers studied roads, automobile engines, locomotives and railroad equipment, electrical generating machinery, sewage collection and treatment, and construction materials.[36]

But few other stations enjoyed such success. A study in the late 1920s found that of 151 colleges conducting some type of research, only 29 spent more than $5,000 a year, and the U.S. Office of Education concluded that "fewer than ten of the land-grant college engineering experiment stations are receiving support from any source sufficient to develop research in engineering." Just two electrical corporations spent twice as much as all college engineering researchers combined. Moreover, much "research" was actually routine testing. Everyone agreed that this work benefited small companies, but many faculty and administrators did not accept Potter's vision of industrial research centers in universities.[37] Realists concluded that engineering experiment stations could not compete against corporate research facilities with multi-million dollar budgets. William Wickenden, author of an evaluation of engineering education in the late 1920s, concluded "it is safe to assume that corporations at this end of the scale are not going to turn to the engineering colleges to get their fundamental problems solved. For one simple reason, they cannot wait long enough."[38]

For that reason, most land-grant schools felt it far more important to connect research to undergraduate education. Certainly engineering faculty thought that the real reason for conducting industrially sponsored research on campus was to show students that research was important and to teach them how to approach research activities. One professor explained in 1917, "The primary function of the university in research should be the training of research men." Industry, he believed, should propose problems whose solutions improved the education of future researchers. "The university should do research primarily to train men and the industry to ensure dividends to its stockholders."[39] A decade later, H. M. Crane, technical assistant to the president of General Motors, put it more bluntly:

> . . . if [colleges] will go to the industry and say, not that they will solve problems,
> but that they will educate the men who, in the future, after they have had training

in the field in addition to the training the university has given them, will solve the problems, they will get all the money they need for research work on an educational basis. In my opinion, that is the only basis on which research work will be a permanent success in the universities.[40]

This logic kept the stations from becoming giant research centers. By 1931 forty stations were in existence, but most were ill-supported. The Illinois station had published 237 bulletins by 1930 and had a budget of $246,000. But much more typical were the stations at Ohio State, Washington State, and Texas A&M, which had annual budgets of about $10,000.[41] Limited resources reinforced the tendency to conduct practical research generally connected to educational and service goals. High on many stations' lists of projects were routine tests of materials, studies of highways problems, and water and sewage treatment.

Clearly, the land-grant schools had found a way to fit research into the engineering program at most schools. But this work was driven by a different imperative than that which motivated German-style research universities like Johns Hopkins and Chicago.[42] At land-grant schools, the ideal of advancing knowledge was clearly secondary to finding answers to immediate problems. Just as land-grant engineering education was oriented toward practice, so research work always kept the real world in view. Some of the work might not have been exciting—attic fan testing at Texas A&M, for example. But this type of research—or testing—was consistent with the overall mission of the universities. Symbolic of this land-grant service ethos in both teaching and research was the creation of a Public Service Engineering program at Purdue in 1935. Open to the best engineering students, the program prepared engineers for government service, an orientation congenial with the original goals of engineering experiment station supporters.[43]

Land-grant schools made one additional lasting contribution to engineering education before the Depression, and that was the establishment of private, incorporated research foundations. Today, these can be found at almost every university. But in 1925, the University of Wisconsin created the first, the Wisconsin Alumni Research Foundation (WARF), to administer agricultural chemist Harry Steenbock's patented process for synthesizing vitamin D. After the regents refused to administer the patent, Steenbock formed a non-profit corporation, "capitalized by privately subscribed funds, managed by 'friends' of the University but operated independently."[44] This patent brought immediate returns to the foundation, which turned the profits back to the university after 1928. By 1950, WARF had given $4.43 million to the University of Wisconsin to support research.[45]

Several schools imitated Wisconsin, but none had the windfall of Steenbock's patent. At Purdue, industrialist and board president David Ross and A. A. Potter established a private foundation to solicit funds, negotiate contracts with individual companies, and sign contracts protecting proprietary information. As one participant explained, "The Purdue Research Foundation was organized . . . to include functions not clearly provided for in the Federal and State laws governing the organization and operation of Purdue University, and so that the rights of each group might be securely protected and an equitable distribution of the profits resulting from cooperative effort guaranteed."[46] The board created the Purdue Research Foundation in 1930.[47] In 1936, Ohio State also formed a research foundation. Prominent industrial leaders, including James F. Lincoln of the Lincoln Electric Company and Charles F. Kettering, head of research of General Motors, again played key roles. They visited Purdue in 1934 and modeled the Ohio State foundation on the Purdue example.[48] By 1940, Minnesota, Cornell, Virginia Tech, Delaware, and Washington State had created research foundations, and Maine was about to. Georgia Tech and Texas A&M acted after the war.[49] In every instance, the projects supported through the foundations continued to be practical in nature, with immediate utility. At Ohio State, for example, foundation projects included confidential work for Drackett Chemical Company (maker of Windex and Drano), and a study of refractories for the Cambria Clay Products Company.[50] Except for sponsorship by a single company, the research itself continued existing patterns at experiment stations.

This continuity in both research activities and educational approaches did not mean that change was absent in the curriculum, the classroom, or the laboratory of the land-grant engineering schools. As scientific understanding led to entirely new areas of engineering, such as radio, the knowledge base steadily shifted toward more science and math. But compared to the best practice in Europe, and even the California Institute of Technology, which overhauled its curriculum completely in the 1920s, the land-grant schools moved quite slowly.[51] Always, land-grant schools maintained a sense of balancing the real world and the classroom; always, practical utility was in sight, whether it took the form of engineering graduates trained for the world of engineering practice, or research aimed at industrial and governmental problems.

Changing Patterns: 1940–1980

World War II marked the beginning of far-reaching changes in engineering education in American universities. For a couple of decades, a different approach to engineering education had been emerging. Robert Thurston had pressed for it at Cornell in the 1890s, but had not influenced most other schools. The high-tech fields, such as electrical and chemical engineering, had demanded more science of their graduates. After 1920, a wave of immigrant European engineers had been demonstrating the utility of applied mathematics and a deeper base of engineering science in engineering curricula. The University of Michigan Summer School in Mechanics, sponsored by Stephen Timoshenko from 1929 through 1936, introduced the new way of thinking about engineering to a generation of graduate students and faculty. But these approaches penetrated American engineering schools, especially land-grant institutions, slowly, and only California Institute of Technology and, to a lesser extent, MIT reconfigured their curricula.[52]

After 1945, however, best practice in engineering education meant placing an emphasis upon engineering science and more rigorous mathematical analysis. A corresponding change in research saw the replacement of industrial research sponsors by federal agencies—especially the military and the Atomic Energy Commission. And the new patrons were interested in research projects that were theoretically inclined toward the cutting edge of technology, with much less emphasis upon practical work, testing, and traditional parameter variation. Projects were larger, more expensive, required much more elaborate research apparatus (which the government also paid for), and relied upon faculty with doctorates supervising graduate students earning doctorates. Topics included jet propulsion, new materials, computers, atomic energy, rockets, and other high-tech arenas.[53] James Kip Finch, dean of engineering at Columbia, summarized the change underway when he explained that "a highly practical and effective program of study, which has emphasized technical methods of use and application, [gave way] to new plans which must have far greater emphasis to the development of basic theory, to more thorough scientific education, and to research and education for research."[54]

Almost overnight, engineering curricula and research enterprises began to reflect the new goals and expectations. Most land-grant engineering colleges were not well-prepared to quickly adjust their programs. Texas A&M, for example, continued to conduct traditional studies of cottonseed oil and attic fans, while providing a very traditional engineering education. As a result, the school managed to attract very little federal research funding.

Illinois, on the other hand, quickly rearranged its entire program of research, and teaching changes followed. On the undergraduate side, a new major in engineering physics was formed to better prepare students for research work in engineering science, while the Department of Theoretical and Applied Mechanics, along with almost every other department in the College of Engineering, moved to expand doctorate programs. Researchers shifted emphasis so rapidly that the flagship projects of the 1930s—home heating and ventilation studies, rail and car wheel behavior, metal fatigue testing, and materials testing for utilities—received almost no notice after the war. Instead, projects funded by the Office of Naval Research (ONR) held the limelight. Illinois engineers had received advance word in late 1944 that ONR intended to fund research after the war, leading the engineering dean and his department heads to develop a strategy in 1945 to win navy contracts. Their successful proposals continued for almost fifteen years, bringing an avalanche of federal money. While industrial research contracts at Illinois totaled $150,000 in 1946, government projects in the College of Engineering alone provided $1,100,000. And instead of individual projects in the $10,000/year range, government work brought projects in the $100,000/year range and up. These included a betatron for the physics department, advanced studies of structural behavior in the Department of Theoretical and Applied Mechanics, the behavior of fluid streams in chemical engineering, and jet propulsion in aeronautical engineering. Other engineers applied theory to welding or studied traveling wave tubes or computers.[55]

In addition, military contracts also provided new facilities for basic research, such as the Control Systems Laboratory opened in 1951 at the request of the Defense Department. The head of the Physics Department oversaw this classified research program that totaled $2.7 million during its first two years of operation. Reorganized as the Coordinated Sciences Laboratory in 1959, the lab remained heavily oriented toward government contracts in computers, plasma, atmospheric and surface physics, and space science. Other research programs at Illinois included a radio telescope, funded in 1958 by the ONR ($323,000), and a nuclear reactor, partially supported by the Atomic Energy Commission in 1958 ($150,000).[56]

Purdue and Ohio State, on the other hand, while taking on some new projects for the military, also retained many of the older, practical projects for industry. Highway materials remained very important subjects for study at both schools. At both schools, trade associations and state government agencies continued to provide most of the research funds. As a result, the total research budgets remained well below the levels at Illinois. But at Georgia Tech, the impact of the new approach to research was even clearer. Engineers at the school had conducted little research before the war, giving some attention to practical studies of turpentine, cotton, and other natural resources. After 1950, however, Georgia Tech became one of the fastest-growing academic research centers in the country, with much work on radar, helicopters, high-speed aerodynamics, microwave acoustics, ultra-high frequency interference, underwater acoustics, and cosmic radiation. Not all of this research was as theoretical as it sounded, for engineers knew to label their work to enhance success with funding agencies. Even so, a school without a research tradition transformed itself using new sources of funding for a different style of research project.[57]

By the late 1950s, most big land-grant schools had made the transition to the new style of government-funded research. They simply could not afford to stay with the prewar pattern. Penn State, for example, gained a multi-million dollar circulating water tunnel as part of a navy research contract, while several universities built wind tunnels, telescopes, large computers, and nuclear reactors. National statistics indicated what federal funding meant to researchers. In 1940, 32 engineering experiment stations had spent a total of $1 million on research. By 1955–56, a smaller number of stations spent $8.2 million from military sponsors, $847,000 from other federal agencies, and $2 million from industry.[58]

Obviously, government replaced industry as the primary supporter of research. The contacts accepted by the Ohio State Research Foundation tell that story. In 1945, 54 industrial and 23 government contracts were underway at Ohio

State, while 29 industrial and 102 government contracts were in force in 1955.[59] But this shift altered a basic relationship that had long distinguished land-grant engineering education. As Courtland Perkins, president of the National Academy of Engineering, explained in 1980, "Industry has disappeared as a major factor in the support of university laboratories. . . . Industry support of university engineering research has been difficult to rationalize in the face of large and multiple support possibilities from government sources." Moreover, graduate students trained using government rather than industrial funds "were not motivated towards the creative interests so important to industry. As a result, the close interplay between industrial researchers and university faculties has dwindled away."[60]

The loss of this close interplay showed in ways large and small. First, it meant that the practice-oriented projects of the prewar era were out. Penn State even dropped the name "engineering experiment station," lest anyone get the wrong idea about the kind of research conducted at its College of Engineering. Other schools repeated this action, and the 46 stations operating in 1946 had dropped to only 30 in 1960. Similarly, educational activities also changed in ways that eliminated certain patterns that had marked the land-grant schools as practice-oriented. Graduate programs became ever more important, while some aspects of undergraduate classes went away. Much less time was spent in the shop-oriented laboratories, just as drafting and surveying gave way to more math and science classes. New majors appeared. Cornell created a major in engineering science in 1946, following on the heels of Stanford. Penn State created an honors course in engineering science in 1953–54, and by 1959 a number of other schools soon offered such curricula; at least seven schools offered engineering physics majors. All limited these programs to their brightest students, those who intended to follow research careers.[61] The perfect symbol of all the changes came in 1960, when Purdue abolished its summer surveying camp, a hallmark of its civil engineering program since 1914.[62] Ironically, in 1962, the Association of Land-Grant Universities merged with its state universities rival, to form the National Association of State Universities and Land-grant Colleges (NASULGC). Here was the bureaucratic evidence that the differences between the two types of school had largely disappeared.

Conclusion: Recent Reforms and the Land-Grant Tradition

In the late 1970s, the first serious complaints about the impact of this postwar style of engineering education and research began to be heard. Many employers began to complain that engineering graduates were completely unprepared to take on normal engineering work; others worried about the almost complete lack of design in curricula. Two engineering educators from the University of Delaware concluded in a 1987 essay in MIT's *Technology Review* that "Design has fallen so low in the order of educational priorities that many engineers—especially young ones and students—do not understand its meaning."[63] Indeed, in 1975, MIT had sponsored a conference to discuss how design could be brought back into the classroom, thus reintroducing the "art of engineering" to students.[64] In general, students have been much better prepared to analyze (usually in mathematical terms) a problem than to solve it in practical terms. The isolated complaints became a chorus of worries in the 1980s and a flood in the 1990s.[65]

At the same time engineering education was being challenged, federal funding began to slip slightly in its importance as a source of research support in engineering colleges—or to be more precise, the focus of those federal funds began to move. Largely undefined basic research projects from the military declined, just as NASA's funding also began to slip. Both changes coincided with much discussion in the late 1970s and 1980s about an apparent decline in American industrial competitiveness. Courtland Perkins was not the only commentator to point to the erosion of ties between engineering schools and industry as a potentially significant factor in this situation.

The responses from engineering schools to these situations have been rather dramatic since the late 1980s. Intriguingly, some of the solutions revive elements of the older land-grant approach to engineering education and research. On the teaching front, a variety of changes have appeared that are nicely captured by the following account from *IEEE Spectrum*.

> On their first day at Drexel University, in Philadelphia, engineering students are ushered into a large auditorium—but not just to sit there passively and listen to a welcoming speech. Instead, Robert Quinn, a professor of electrical and computer engineering, teams the freshmen up in threes and instructs each team to design a model bridge using a toy construction set called Connects. . . . From then on, these students will have to devote as much attention to building the skills they will need in the new team-oriented, multidisciplinary industrial environment as to learning differential calculus and circuit analysis.[66]

Teamwork, problem-solving, hands-on experience—all of these are part of the effort to "re-engineer engineering education," as the author of that *Spectrum* article phrased it.

These adjustments, and others as well, are reflected in the first substantial revision in several decades of the accreditation standards used by the Accreditation Board for Engineering and Technology (ABET). These curricular standards no longer adopt a "Chinese menu" approach of specifying so many courses from various areas of study. Instead, they emphasize the development of "competencies" that are a direct reflection of the needs and demands of engineering practice. This approach allows much greater curricular flexibility for academic departments, but it is also a direct response to the criticism that engineering college graduates could not solve real-world problems.[67]

Nor were the changes limited to undergraduate education. In 1995, MIT president Charles James Vest announced a new master's of science in engineering degree that "has more of a practice than a research orientation." Designed for students bound for industry, this is a non-thesis degree intended to be completed eighteen months faster than a traditional research MS. Vest suggested that the Institute planned to place more emphasis on such master's-level work, and was intending to develop integrated bachelor's-master's programs. "We have to expand the horizons of our students so that they have a better sense of the variety of careers out there . . . beyond being a faculty member or working at the most advanced levels in a research organization."[68] A number of schools have been following suit, recognizing that the traditional pattern of protracted study served neither the student nor the sponsor. But for research-oriented MIT, this was a significant development that marked a break with fifty years of expectations that only the doctorate really mattered.

This shift in graduate programs was paralleled by an adjustment on the research side, as many engineering colleges, land-grant and others alike, consciously set out to revive ties to industry. At the University of Wisconsin, for example, a program launched in 1963 (University-Industry Research Program) but which had never attracted much attention, provided the basis for building much deeper connections with industrial research sponsors during the 1980s. In other states, legislatures supported programs intended to link university researchers and local industries, such as the Ben Franklin Partnership in Pennsylvania (1983) and the Thomas Edison Program in New Jersey.[69] Purdue established two programs, one to promote technical education and the other to link university engineers and industries needing research assistance. MIT's Industrial Liaison Program celebrated fifty years of existence in 1998, but it was clearly less important than research programs connected to Washington for most of those years.[70]

The impact of such initiatives became apparent in the annual accounting of research funding at the nation's engineering schools. Engineering college research expenditures in 1982–83 had included more than $761 million in federal funds, while industry provided $141 million. The latter number represented an increase from $119 million the previous year, but by 1989, industrial funding had jumped to $850 million.[71] Biotechnology and microelectronics have been especially trendy topics, but across the board, industrial research has regained much of the prestige it lost in the aftermath of World War II.

These readjustments seem to be reviving pieces of the historical patterns of land-grant engineering education, but not just in land-grant universities. Despite rhetorical flourishes like the comment of MIT's James Killian in 1980 that "cooperative research is an idea whose time has clearly come," recent research programs bear obvious resemblances to the traditional efforts that distinguished engineering experiment stations.[72] Thus efforts to attract state seed money or corporate research funds are buttressed with arguments resembling those made by the stations at the start of the century. Connecting engineering research to regional economic growth, or finding ways to serve industry through education or research, is not new at all. It is the scale of the effort that is different, and the research approaches also have changed. In North Carolina, for example, a $30 million microelectronics research facility at the Research Triangle Park and a $43 million Microelectronics Center demonstrated the new way of doing things. Pennsylvania's Ben Franklin Partnership funded several advanced technology centers at universities in the state, notably Penn State. The Georgia Tech Research Institute, successor to the engineering experiment station, sponsored twelve field offices to increase contacts with firms with fewer than 100 employees. One field office director noted that "We provide [brief managerial and technical problem-solving] services at no cost for three to five man-days. Beyond that, the institute charges to recover its own costs."[73]

All this is not to suggest that history is a circular enterprise. Indeed, there are substantial differences between the engineering colleges of the 1920s and those of the twenty-first century. Land-grant colleges are no longer sharply separated, as they once were, by unique curricula, service philosophies, or academic research programs. For good reason, the older hands-on approach to education would no longer work with many of today's engineering and technology fields. But the land-grant schools made substantial contributions to the development of ways of thinking and doing within engineering education that served the nation well in the past. The recognition that engineering is not simply an academic exercise, but one that takes place in the real world was a pivotal aspect of the land-grant engineering tradition, and it turns out to remain viable still in the changing engineering world of the twenty-first century.

Notes

1. See Terry S. Reynolds, "The Education of Engineers in America Before the Morrill Act of 1862," *History of Education Quarterly* 32 (Winter 1992): 459–82; O. Allan Gianniny, Jr., "The Overlooked Southern Approach to Engineering Education: One and a Half Centuries at the University of Virginia, 1836–1986," in Howard L. Hartman, ed., *Proceedings of the 150th Anniversary Symposium on Technology and Society: Southern Technology: Past, Present and Future*, College of Engineering, University of Alabama, Tuscaloosa, March 3–4, 1988. See also Lawrence P. Grayson, "A Brief History of Engineering Education," *Engineering Education* 67 (1977): 246–64; and James G. McGivern, *First Hundred Years of Engineering Education in the United States (1807–1907)* (Spokane, WA: Gonzaga University Press, 1960).

2. The term "land-grant" came from the provision in the legislation that awarded each state 30,000 acres of land for every congressman and senator. The classic histories of land-grant colleges are Edward D. Eddy, Jr., *Colleges for Our Land and Times: The Land-Grant Idea in American Education* (New York: Harper & Brothers, 1957); Earle D. Ross, *Democracy's College: The Land-Grant Movement in the Formative Stage* (Ames: Iowa State University Press, 1942); see also Frederick Rudolph, *The American College and University: A History* (New York, Vintage Books, 1962).

3. W.E. Dalby, "The Training of Engineers in the United States," *Proceedings of the Institute of Naval Architects* 45 (1903): 39.

4. This issue is the paramount theme of the classic study by Monte A. Calvert, *The Mechanical Engineer in America, 1830–1910: Professional Cultures in Conflict* (Baltimore: The Johns Hopkins University Press, 1967).

5. Ibid., p. 48.

6. See Henry C. Dethloff, *A Centennial History of Texas A&M University, 1876–1976,* 2 vols. (College Station: Texas A&M University Press, 1975).

7. University of Illinois, *Report of the Board of Trustees* (1870–1871), p. 41, quoted in Ira O. Baker and Everett E. King, *History of the College of Engineering of the University of Illinois, 1868–1945* (Urbana, typescript copy in University archives, c. 1946), p. 242.

8. Baker and King, *A History of the College of Engineering,* p. 83. The following material on Robinson is drawn from this source.

9. Information on Robinson from Baker and King, *History,* pp. 83, 198–204, 369–70.

10. See Papers of Anson Marston, Iowa State University Archives, Ames, Iowa; also Herbert J. Gilkey, *Anson Marston: Iowa State University's First Dean of Engineering* (Ames, 1968), pp. 12–15, 18.

11. Information about the technical institutes can be found in Robert Thurston, "Technical Education in the United States," *Transactions of the American Society of Mechanical Engineers* (1893): 931–33; Herbert Foster Taylor, *Seventy Years of Worcester Polytechnic Institute* (Worcester: Davis Press, Inc., 1937); Mildred McClary Tymeson, *Two Towers: The Story of Worcester Tech, 1865–1965* (Barre, MA: Barre Publishers, 1965); Calvert, *Mechanical Engineer in America,* p. 49; and Stevens Institute of Technology, *Stevens 75th Anniversary, Commemorating 75 Years of Accomplishment in Engineering Education* (Hoboken, NJ: Alumni Association of Stevens Institute of Technology, 1945).

12. Baker and King, *History,* p. 162.

13. Harris J. Ryan, Analysis of candidate for Head of EE, Folder 22: Electrical Engineering 1927–1932, Box 2, Series 1, School of Engineering Records, SC 165, Special Collections, Stanford University Archives, Palo Alto, CA.

14. See Calvert, *The Mechanical Engineer,* pp. 45–57; Robert H. Thurston, *The Mechanical Engineer: His Preparation and Work; An Address to the Graduating Class of the Stevens Institute of Technology* (New York, 1875); and William F. Durand, *Robert Henry Thurston; A Biography, The Record of a Life of Achievement as Engineer, Educator, and Author* (New York: ASME, 1929).

15. Quotation from Calvert, *Mechanical Engineer in America,* p. 47; see also Anson Marston, "Original Investigations by Engineering Schools a Duty to the Public and to the Profession," *Proceedings of the Society for the Promotion of Engineering Education* 8 (1900): 237; and Cornell University, *Annual Report of the President* (1896–1897), pp. xl–xli; (1897–1898), pp. 42–43.

16. Maurice Caullery, *University and Scientific Life in the United States,* trans. James Hamilton and Emmet Russell, (Cambridge, MA: Harvard University Press, 1922), pp. 121–22; quotation in the text from an article by George Swain, engineering professor at Harvard, from *Science* (January 2, 1910): 81–93.

17. Quotation from H. H. Higbie, "Research in Engineering Colleges of Interest to Industry," *Journal of Engineering Education* 23 (October 1932): 154. Similar concerns were documented by a massive study of engineering education during the 1920s, usually labeled the Wickenden Report. Society for the Promotion of Engineering Education, *Report of the Investigation of Engineering Education, 1923–1929*, 2 vols. (Pittsburgh: SPEE, 1930, 1934); see also Terry S. Reynolds and Bruce E. Seely, "Striving for Balance: A Hundred Years of the American Society for Engineering Education," *Engineering Education* 82 (July 1993): 138–40.

18. Embury A. Hitchcock, *My Fifty Years in Engineering: The Autobiography of a Human Engineer* (Caldwell, Idaho: Caxton Printers, 1939), pp. 75–78, 91–112.

19. Information on agricultural experiment stations from A. Hunter Dupree, *Science in the Federal Government: A History of Policies and Activities to 1940* (Cambridge: Belknap Press, 1957), pp. 149–83; T. Swann Harding, *Two Blades of Grass: A History of Scientific Development in the U.S. Department of Agriculture* (Norman: University of Oklahoma Press, 1947), pp. 173–92; and Alan I Marcus, *Agricultural Science and the Quest for Legitimacy: Farmers, Agricultural Colleges, and Experiment Stations* (Ames: Iowa State University Press, 1985); and several essays by Charles E. Rosenberg in his collection, *No Other Gods: On Science and American Social Thought* (Baltimore: Johns Hopkins University Press, 1990 [1976]).

20. "WARF Funds Play Vital Role in UW Research Activities," *UIR Memo* 1, no. 3 (1965): 5. (*UIR Memo* was a publication of the University-Industry Relations Office at the University of Wisconsin.)

21. Gilkey, *Anson Marston*, pp. 13–15, 21–25.

22. The pattern of faculty and industrial consulting is found in the histories of engineering schools and in autobiographies. For examples, see J. Merrill Weed, "The Second Quarter Century of the College of Engineering at the Ohio State University," Part 1, *Centennial History of the Ohio State University* (1969), p. 7; Baker and King, *College of Engineering of Illinois*; H. B. Knoll, *The Story of Purdue Engineering* (West Lafayette: Purdue University Studies, 1963); Hitchcock, *Fifty Years in Engineering*; and Gilkey, *Anson Marston*, pp. 13–15. On Jackson, see W. Bernard Carlson, "Academic Entrepreneurship and Engineering Education: Dugald C. Jackson and the MIT-GE Cooperative Engineering Course, 1907–1932," *Technology and Culture* 29 (July 1988): 543–4.

23. See "Arthur Newell Talbot Laboratory," *University of Illinois Bulletin* 35, no. 62 (April 1, 1938), pp. 37–38; Baker and King, *History*.

24. Baker and King, *College of Engineering of Illinois*, p. 118; also Knoll, *Purdue Engineering*, pp. 191–93.

25. "History of the Department of Ceramic Engineering," pp. 1–7, in Ohio State University, *Centennial History of the College of Engineering*, part 2; and Arthur S. Watts, "Ceramic Engineering," pp. 45–49, in U.S. Department of the Interior, Bureau of Education, *Land-Grant College Education, 1910–1920*, Part IV, *Engineering and the Mechanic Arts*, Bulletin no. 5 (1925).

26. Efforts to gain federal research funds for engineering experiment stations peaked as the United States entered World War I, but funds never materialized, largely be-

cause members of Congress concluded that adequate resources for industrial and engineering research existed. See several talks and other information in the Papers of Anson Marston, Iowa State University Archives, Ames, Iowa; Willis R. Whitney, "Engineering Experiment Stations in State Colleges," *Science* 43 (June 23, 1916): 890–91, 895–96; Henry H. Armsby, "A Review of Proposals for Federal Support of Engineering Research in the Colleges," *Engineering Experiment Station Record* 27 (October 1947): 39; and Daniel J. Kevles, "Federal Legislation for Engineering Experiment Stations," *Technology and Culture* 12 (April 1971): 112–19.

27. Anson Marston, "Original Investigations by Engineering Schools a Duty to the Public and the Profession," *Proceedings of the Society for the Promotion of Engineering Education* 8 (1900): 244.

28. A. N. Talbot, "Tests of Reinforced Concrete Beams," *University of Illinois Engineering Experiment Station Bulletin* no. 1 (1904), cover. For more information on the station's founding, see L. P. Breckinridge, "The Engineering Experiment Station of the University of Illinois," *University of Illinois Engineering Experiment Station Bulletin* no. 3 (March 1906); Baker and King, *History of the College of Engineering*, pp. 767–833.

29. The charter was printed inside every *Bulletin* of the Iowa State Engineering Experiment Station during its early years.

30. Information drawn from the annual compilations of the *Engineering Experiment Station Record* during the 1920s.

31. Ellery B. Paine, "The Engineering Experiment Station of the University of Illinois," *Proceedings of the American Institute of Electrical Engineers* 34 (October 1915): 2421.

32. Arthur N. Talbot, "New Work for Testing Society," *Iron Trade Review* 55 (July 2, 1914): 16.

33. Information from the project files of the Illinois Engineering Experiment Station, University Archives, University of Illinois, Urbana.

34. Information in E. A. Hitchcock and J. M. Weed, *A Description of the Engineering Experiment Station, Ohio State University*, Bulletin No. 50, Ohio State University Engineering Experiment Station (1929); articles in *Engineering Experiment Station News*, published quarterly after 1929 by the station at Ohio State; *Engineering Experiment Station Record*; A. A. Potter and G. A. Young, "Tendencies in Research at Engineering Colleges," *Society of Automotive Engineers Journal* 20 (May 1927): 623–32; A. A. Potter, "Research Relations Between Colleges and Industry," *American Institute of Electrical Engineers Journal* 45 (1926): 1272–76; and R. A. Seaton, "How Industry Can Cooperate with Engineering Colleges in Furthering Research," *Journal of Engineering Education* 16 (November 1925): pp. 203–14.

35. Purdue University, Engineering Extension Department, *Proceedings of Industrial Conference held at Purdue University, June 1, 1926*, Bulletin no. 15 (June 1926), pp. 12–13.

36. Figures from the Association of Land-Grant Colleges, *Proceedings of the Annual Convention*, (1920–31); "Purdue as an Engineering Center," *Domestic Engineering* 124 (July 7, 1928): 18–20; and Purdue University, Engineering Extension Department, *Proceedings of Industrial Conference held at Purdue University, June 1, 1926*,

Bulletin no. 15 (1926), pp. 24, 22. On Potter, see David F. Noble, *America by Design: Science, Technology, and the Rise of Corporate Capitalism* (New York: Knopf, 1977), pp. 180–81. On other aspects of Purdue, see Knoll, *Purdue Engineering;* and "Twenty-Five Years of Progress," *Purdue Research Foundation Horizon* 2 (May 1956), pp. 2–3.

37. Potter was not alone in his view. See Benjamin F. Bailey, "Can the University Aid Industry?" *Journal of the American Institute of Electrical Engineers* 45 (1926): 742–745; P.M. Heldt, "Industrial Research—Where the College Fits In," *Automotive Industries* 56 (March 5, 1927): 355–356; Seaton, "How Can Industry Cooperate"; and A. A. Potter, "Cooperative Engineering Research," *Journal of Engineering Education* 22 (October 1931): 92–4.

38. William E. Wickenden, "Research in the Engineering Colleges," *Mechanical Engineering* 51 (August 1929): 587.

39. C. E. Skinner, "Industrial Research and its Relation to University and Governmental Research," *Proceedings of the American Institute of Electrical Engineers* 36 (October 1917): 767, 770.

40. Comment to A. A. Potter and G. A. Young, "Tendencies in Research at Engineering Colleges," *SAE Journal* 20 (May 1927): 632. For a similar remark, see J. B. Whitehead, "Industry and Engineering Colleges," *Journal of Engineering Education* 16 (November 1925): 218–223.

41. Information from *Engineering Experiment Station Record* (1923–1931).

42. This point of view is best developed in Roger L. Geiger, *To Advance Knowledge: The Growth of American Research Universities, 1900–1940* (New York: Oxford University Press, 1986).

43. Knoll, *Story of Purdue Engineering,* p. 99; and *Engineering Experiment Station News* (OSU), 7 (February 1935): 8–9.

44. "WARF Plays Vital Role," *UIR Memo* 1, no. 3 (1965): 7.

45. For the period 1950–1973, WARF provided $50 million to the university. See information in ibid.; and in Allan G. Bogue and Robert Taylor, eds., *The University of Wisconsin: One Hundred and Twenty-Five Years* (Madison: The University of Wisconsin Press, 1975).

46. G. S. Meikle, "The Department of Research Relations with Industry," in Purdue University, *Annual Report of the President,* 61 (1935): p. 164. Meikle was director of the new Department of Research Relations with Industry.

47. Information from Knoll, *Purdue Engineering,* pp. 94–96; "Twenty-Five Years of Progress," pp. 2–3; Purdue University, *Annual Report of the President* 60–65 (1932–33 to 1939–40); Purdue Research Foundation, *Purdue Research Foundation: Its Organization and Purpose,* Bulletin no. 1 (1931); idem, *Proceedings of the Second Research Conference, October 9 and 10, 1931,* Bulletin no. 2 (1931).

48. Robert C. Thompson, "A History of the Ohio State University Research Foundation, 1936–1969" (September 1969), pp. 1–21, prepared for the Ohio State University Centennial, copy in Ohio State University Archives, Columbus, Ohio.

49. A. A. Potter, "Research and Invention in Engineering Colleges," *Science* 91 (January 5, 1940): 1–7; and Cornelia Weil, "UDRF: Seed Money for Science," *University of Delaware Magazine* (Fall 1988): 12–15.

50. The Ohio State University Research Foundation, *Annual Report* 1–3 (1937–1939), copies in Research Foundation: Annual Reports, Box 1, RG 38/c/1, The Ohio State University Archives; also Thompson, "History of the Ohio State University Research Foundation," pp. 18–21.

51. I discuss these changes in "The Other Re-engineering of Engineering Education, 1900–1965," *Journal of Engineering Education* 88, no. 3 (July 1999): 285–94; and in "Research, Engineering, and Science in American Engineering Colleges, 1900–1960," *Technology and Culture* 34 (April 1993): 344–86.

52. I deal with these changes at length in "The Other Re-engineering of Engineering Education, 1900–1965," *Journal of Engineering Education* 88, no. 3 (July 1999): 285–94; and "Research, Engineering, and Science in American Engineering Colleges, 1900–1960," *Technology and Culture* 34 (April 1993): 344–86.

53. A number of studies deal with the changes in postwar research and education. See Daniel J. Kevles, *The Physicists: The History of a Scientific Community in Modern America* (New York: Knopf, 1977); and Stuart W. Leslie, *The Cold War and American Science: The Military-Industrial-Academic Complex at MIT and Stanford* (New York: Columbia University Press, 1993).

54. James Kip Finch, *Trends in Engineering Education: The Columbia Experience* (New York: Columbia University Press, 1948), pp. 5–6.

55. See correspondence in various files related to the ONR for 1945–46, but especially in ONR General Correspondence, 1945–54, Subject File, 1909–61, Box 7, Engineering Experiment Station Records, Series 11/2/1, University of Illinois Archives, Urbana; University of Illinois, *Report of the Board of Trustees*, 44 (1946–48); see especially May 29, 1946, pp. 1103–04. Statistics on general developments from Kevles, *The Physicists*, pp. 354–55.

56. University of Illinois, *Report of the Board of Trustees* 47 (September 23, 1953), pp. 1087–89; "The Coordinated Science Laboratory: Some Historical Highlights," (March 10, 1964), in Engineering—Coordinated Science Laboratory, File 1964–67, Series 11/30/05, University of Illinois Archives. Statistics from University of Illinois, *Report of the Board of Trustees* 50 (1958–60), pp. 190, 572; 49 (1956–58), p. 1240.

57. See Knoll, *Purdue Engineering*; McMath et al., *Engineering the New South*, pp. 212–17, 256–70.

58. Information on Penn State from Michael Bezilla, *Engineering Education at Penn State: A Century in the Land-Grant Tradition* (University Park, PA, 1981); statistics from annual review of engineering graduate study and research, conducted and compiled by the American Society for Engineering Education and printed in *Engineering Education*.

59. Ohio State University Research Foundation, *Annual Reports*, 1937–1955, in RG 38/c/1, Box 1, Ohio State University Archives, Columbus.

60. Courtland D. Perkins, "Public Policy and the Innovation Process," in Nam P. Suh and Bruce M. Kramer, eds., *Cooperative Research: Proceedings of a Conference Spon-*

sored by the Laboratory for Manufacturing and Productivity, Massachusetts Institute of Technology, and the National Science Foundation (Cambridge: The MIT Press, 1982), pp. 40–41.

61. Bezilla, *Engineering Education at Penn State*, pp. 172–74.

62. Knoll, *Purdue Engineering*, pp. 251–54, 260–61.

63. Eugene E. Covert, "Engineering Education in the '90s: Back to Basics," *Aerospace America* 30, no. 4 (April 1992): 21; see also Arnold D. Kerr and R. Byron Pipes, "Why We Need Hands-on Engineering Education," *Technology Review* 90, no. 7 (October 1987): 38.

64. Massachusetts Institute of Technology, Center for Policy Alternatives, *Future Directions for Engineering Education: System Responses to a Changing World* (Washington, DC: ASEE, 1975).

65. See, for example, John R. Dixon, "New Goals for Engineering Education," *Mechanical Engineering* 113 (March 1991): 56–62; Peter J. Denning, "Educating a New Engineer," *Communications of the ACM* 35, no. 12 (December 1992): 82–97; and Gary P. Maul, "Reforming Engineering Education," *Industrial Engineering* 26, no. 10 (October 1994): 53–55, 67; David T. Curry, "Engineering Schools Under Fire," *Machine Design* 63 (October 10, 1991): 50.

66. G.C. Masi, "Re-engineering Engineering Education," *IEEE Spectrum* 32, no. 9 (September 1995): 44.

67. The Engineering Criteria 2000 can be found on the ABET website: http://www.aBET. ORG/downloads/2000_01_Engineering_Criteria.pdf.

68. Charles M. Vest, "U.S. Engineering Education in Transition," *The Bridge* 25 (Winter 1995): 4–9, quotations from p. 8.

69. Information on Wisconsin from *UIR Memo*, Wendt Library, University of Wisconsin; on other schools, see David Dickson, *The New Politics of Science* (Chicago: University of Chicago Press, 1988); and David Osborne, *Laboratories of Democracy*, (Boston: Harvard Business School Press, 1988).

70. Text from MIT's Industrial Liaison Program home page: http://ilp.mit.edu/ilp/General/WhoWeAre.html

71. Most statistics are from *Engineering Education*, annual issue with report on graduate education and research. Other numbers from "Business Goes to College," *Business Week* (July 16, 1989): 50.

72. James Killian, "Introductory Remarks," in Nam P. Suh and Bruce M. Kramer, eds., *Cooperative Research: Proceedings of a Conference Sponsored by the Laboratory for Manufacturing and Productivi.y, Massachusetts Institute of Technology, and the National Science Foundation* (Cambridge: MIT Press, 1982), p. 1.

73. Mark A. Fischetti and John Horgan, "Learning to Satisfy Market Needs," *IEEE Spectrum* 21 (November 1984): 106.

■ Index